资助项目及单位
Funding Project and Organization

国家基金委基础科学人才培养基金（J1030518）
国家级实践教学示范中心建设项目（教高函[2007]21号）
中国地质大学（武汉）设备处
中国地质大学（武汉）教务处

National Science Fund for Talent Training in Basic Science (J1030518)
National Demonstration Center for Practical Teaching ((2007) No.21)
Facilities Management Office of China University of Geosciences (Wuhan)
Academic Administration of China University of Geosciences (Wuhan)

周口店野外实践教学基地
经典地质现象图册

Atlas of Classic Geological Phenomena in the Zhoukoudian Field Practical Teaching Base

主 编 赵俊明　　副主编 袁晏明
Chief Editor: Zhao Junming　　Subeditor: Yuan Yanming

中国地质大学出版社
CHINA UNIVERSITY OF GEOSCIENCES PRESS

图书在版编目(CIP)数据

周口店野外实践教学基地**经典地质现象图册**:汉英对照 / 赵俊明主编 —武汉:中国地质大学出版社,2011.2(2017.7重印)

ISBN 978-7-5625-2494-6

Ⅰ.①周…
Ⅱ.①赵…
Ⅲ.①周口店(考古地名)-区域地质-图集-汉、英
Ⅳ.①P562.13-64

中国版本图书馆 CIP 数据核字(2011)第 013974 号

周口店野外实践教学基地 **经典地质现象图册**	赵俊明 主编	袁晏明 副主编
责任编辑:刘桂涛 李晶		责任校对:戴 莹
出版发行:中国地质大学出版社(武汉市洪山区鲁磨路388号)		邮政编码:430074
电 话:(027)67883511 传 真:(027)67883580		E-mail: cbb@cug.edu.cn
经 销:全国新华书店		http://www.cugp.cug.edu.cn
开本:787毫米×1 092毫米 1/12		字数:460千字 印张:14.75
版次:2011年2月第1版		印次:2017年7月第2次印刷
印刷:荆州鸿盛印务有限公司		印数:3001—4000册
ISBN 978-7-5625-2494-6		定价:138.00元

如有印装质量问题请与印刷厂联系调换

中国地质大学(原北京地质学院)1960级校友、国家总理温家宝在校八年的学习期间,曾五次到周口店实践学习。2004年7月17日,中国地质大学隆重举行周口店实习站建站五十周年庆典活动,总理温家宝同志亲笔题字"摇篮"以示祝贺,是对周口店野外实践教学的充分肯定和殷切期望。

  The prime minister Wen Jiabao, graduated from Beijing Institute of Geology (former name of China University of Geosciences), came five times to Zhoukoudian for field practice and learning during his eight-year education in our university in 1960s. On July 17th, 2004, the 50th anniversary of the establishment of the Base, the prime minister Wen Jiabao autographically inscripted "Cradle" (Yaolan) as a felicitation, which expresses his satisfying affirmation and eager expectation for the Zhoukoudian field practical education.

考古寻宗铭华夏
地质摇篮慰古今

杨遵仪
2004.5.10.

周口店实习站地处西山位近燕同
适逢五十周年惠及学子逾万
赋志致贺

千度晚则百道剑闯师生
修业半世纪
追踪亿万年
地学圣区人类摇篮演变

王鸿祯
二〇〇四年五月

庆祝周口店实习站
建站五十周年

实践出真知
精英的摇篮

杨起 二〇〇四

塑造地质人生
的起点
锻造地质品格
的熔炉

周口店实习站
建站五十周年
纪念

赵鹏大
二〇〇四年五月

地质摇篮

殷鸿福
2004.5.9

院士为周口店实践教学基地题词
Inscriptions to the Zhoukoudian Field Practical Teaching Base by academicians

希望我校周口店地质工作者的摇篮不断培养出世界一流的地学人才。

於崇文
2004.06.01

庆祝中国地质大学周口店教学实习基地建立五十周年

地质英才启蒙的窗口，
地学栋梁培育的摇篮。

祝愿实习基地在科学化、现代化的道路上取得更大的胜利！

张本仁
2004年5月9日

借天地造化之奇
育千万地学英才

莫宣学
二〇一〇年五月

精细观察　见微知著
缜密思考　溯本求源

发扬周口店站的科学
精神和优良传统

翟裕生
二〇〇四年五月

勤能不息。
酬勤不用。
天道酬勤
自强不息
求真致用
学术争先。

金振民
二〇一〇年三月十九日

院士为周口店实践教学基地题词
Inscriptions to the Zhoukoudian Field Practical Teaching Base by academicians

序 一

周口店位于华北板块中部，处于太行山脉和燕山山脉枢纽处，保存了从太古代到新生代的地层序列。漫长的地质历史和多次的地壳运动造就了形态迥异的地貌单元、种类繁多的岩石类型和奇特丰富的构造现象，著名的周口店"北京人遗址"被联合国确定为"世界自然与文化遗产"而享誉世界。

周口店是我国地质研究史和教育史的一片圣地。自1954年被誉为"地质工作者摇篮"的中国地质大学（原北京地质学院）周口店实习基地建站以来，这里培养了数以万计的学生。他们在周口店接受了地质启蒙教育和野外生活的挑战后，奔赴祖国建设的各条战线，从中涌现出了包括党和国家领导人、数十名中国科学院和中国工程院院士、数百名资深地质学家和教育家在内的无数杰出人才，取得了辉煌的成果。

50多年来，在各级领导和师生的共同努力下，经过不断探索、改革和创新，逐步形成了注重理论与实践相结合、敢于探索、一切从严的具有周口店特色的野外实践教学方法，培养了师生们团结协作、艰苦朴素、勇于创新的工作作风。周口店实习基地成为中国地质大学人才建设的培养基地。

近年来，到周口店进行野外地质实践教学的师生人数不断增加，仅中国地质大学（武汉）每年就有近千名师生在此实习，此外还有中国地质大学（北京）、中国石油大学、中国矿业大学、中国科技大学、河南理工学院等兄弟院校的师生来此进行为期不等的地质实践教学，各地质现象观察点显得日益拥挤。同时随着当地对山体资源的逐步开发，许多经典地质现象开始遭到不同程度的破坏，对野外地质实践教学造成了不利影响。为了满足教学需要，使教师和学生可以在野外路线教学之后的室内整理阶段能身临其境般地进行详细讲解和回顾，同时也为了使这些具有历史传承和地质美学意义的地质现象能够被更多的人所认识，在赵俊明老师等的共同努力下，出版了这本《周口店野外实践教学基地经典地质现象图册》。

本书以周口店野外实践教学大纲为框架，分为地层古生物、岩石矿物、构造、第四纪地质及地貌和资源与环境五个部分，囊括了每条教学路线的观察点，同时还对非教学路线上的经典地质现象一一做了收录。该书以精美的图片为骨干，配以详细的文字描述和解说，使读者仿佛置身于周口店的大山中，既欣赏了大自然鬼斧神工般的地质奇观，又掌握了科学知识。同时，对于地层学、构造地质学、岩石学等课程的室内教学来说，本书丰富的实例图片，也对课堂理论讲授起了很好的互补作用。

可以相信，本书的出版将给周口店野外地质实践教学起到极大的帮助，使师生们随时随地可以通过本书，对每条教学路线和各种地质现象进行复习和巩固，达到了将大山搬进课堂的效果。

中国科学院院士
中国地质大学（武汉）教授

殷鸿福
2010年6月18日

PREFACE

Zhoukoudian located at the junction of Taihang Mountain range and Yanshan Mountain range in the central North China Craton, preserves stratigraphic sequences ranging from the Archean to the Cenozoic. Long geological history and several episodes of crustal movements created a wide variety of landscape units, rock types and peculiar tectonic phenomena. The world-famous "Peking Man Site" in Zhoukoudian has been inscribed in the world natural and cultural heritage list by the UNESCO.

Zhoukoudian is a "holy land" in the history of China geological research and education. Tens of thousands of students have been educated since the Zhoukoudian field practice teaching base that is known as the "cradle of geologists" was founded by China University of Geosciences (former Beijing College of Geology) in 1954. With primary geological education and living challenges in the field in Zhoukoudian, the graduates went to various front lines of the construction for our motherland, and achieved outstanding successes. Among them dozens have become academicians of the Chinese Academy of Sciences and the Chinese Academy of Engineering, and hundreds have become senior geologists and educators. A great number of excellent graduates have played active roles as major leaders in the Party and government.

For the past over 50 years, with persistent reform, innovation and exploration, a characteristic teaching method for the field practice in Zhoukoudian has been formed under joint efforts of teachers, students, and leaders at all levels. It[AW1] emphasizes the combination of theory and practice and of exploration and strict training, and trains teachers and students of the work style of unity and cooperation, plain living and innovation. The Zhoukoudian field practice teaching base has become a talent training base of China University of Geosciences.

In recent years, more and more students and teachers come to Zhoukoudian for geological field practice. Nearly a thousand students from the China University of Geosciences (Wuhan) alone come here every year. In addition, students and teachers from other universities such as China University of Geosciences (Beijing), China University of Petroleum, China University of Mining and Technology, University of Science and Technology of China, Henan Polytechnic University also conduct field practice here for different periods. So, teaching sites for geological phenomena are increasingly developed. However, with the local government exploitation of mountain resources step by step, many classic geological phenomena have suffered varying degrees of damage, which led negative impacts on geological field practice teaching. To satisfy teaching needs, allow students and teachers to review field phenomena indoors, and make geological phenomena with historic heritage and geo-esthetic significance known to more people, the "*Atlas of Classic Geological phenomena in the Zhoukoudian Field Practical Teaching Base*" is published with joint efforts of Zhao Junming and other teachers.

Based on the Zhoukoudian field practice teaching program, the atlas mainly consists of five parts: Stratigraphy and Paleontology, Rocks and Minerals, Structural Geology, Quaternary Geology and Geomorphology, Resources and Environment, covering not only observation sites of each teaching route, but also geological phenomena of non-teaching routes. With colorful pictures as the main body, which are accompanied by detailed text descriptions and explanations, the atlas brings readers feeling as if being in the mountains of Zhoukoudian, and appreciation to geological wonders created by nature while mastering scientific knowledge. In addition, the atlas is rich in example pictures and is a nice complement to theoretical teaching indoors for subjects like Stratigraphy, Structural Geology and Petrology.

I believe that the publication of the atlas would be of great help to Zhoukoudian geological field practice teaching. Teachers and students can review and consolidate understanding of all of the geological phenomena in each teaching route anytime and anywhere, which accomplishes the effect of "bringing the mountains into the classroom".

Academician of Chinese Academy of Sciences
Professor of China University of Geosciences (Wuhan)

Yin Hongfu
June 18, 2010

序 二

　　周口店野外实践教学是中国地质大学最具传承、最有特色的教学内容。周口店地区特殊的地理位置与地质构造单元，成就这里具有得天独厚的地质现象与地理景观，岩类齐全，构造现象丰富，华北地区寒武纪以来的典型地层稳定，在时间上、空间上的地质演化与发展极具代表性，是我国地质类基础教学难得的野外实践基地。20世纪50年代初，前苏联地质学家帕夫林洛夫、中科院院士池际尚、马杏垣等老一辈地质学家在周口店开展野外教学实习，拉开了北京地质学院（中国地质大学前身）周口店实习基地建设的序幕。50多年来，超过4万名学生在此参加过野外实习，为国家输送了包括国家领导人、数十名两院院士、国家优秀登山运动员等在内的无数杰出人才。只要参加过周口店实习的每个地大毕业生，都会在心中对周口店实习留下深深的印迹，并成为一生中最值得怀念的一段时光。2004年周口店实习基地庆祝建站50周年之际，曾经五次在周口店地区实习的校友温家宝总理欣然题写了"摇篮"两字，以表彰周口店实习站为中国地质事业做出的贡献。

　　周口店野外教学具有实践性强、理论知识与野外地质现象之间具有很强对应性等特点，对学生牢固掌握野外地质技能、学习巩固专业知识具有很好的效果。因此，自建站以来一直受到学校的高度重视。中科院院士杨遵仪、王鸿祯、赵鹏大、殷鸿福、於崇文、金振民等都曾亲临周口店授课，学校领导每年都去实习站关心与指导。经过一代又一代地质学家多年的摸索总结，形成了严谨探索、循序渐进、反复实践与认识的教学方法，培养了学生艰苦朴素，团结协作，勇于创新的思想作风与工作作风。

　　进入21世纪，如何切实保障野外实践教学质量，不断提高野外实践教学水平，培养适应现代地质工作的野外基本工作技能，成为我校周口店野外教学实践工作的重中之重。为了进一步丰富周口店野外地质教学资源，为教师、学生提供更为典型、直观和鲜明的图片资料，以赵俊明教授为负责的课题组，历时三年，终于将《周口店野外实践教学基地经典地质现象图册》在中国地质大学建校60周年暨周口店建站58周年前夕呈现给大家。按照周口店野外实践教学大纲的要求，该图册将周口店经典地质现象一一整理成册，以精美的野外照片为主，分为地层古生物、构造、岩石矿物、第四纪地貌以及资源与环境五个部分，涵盖了包括所有教学路线在内的经典地质现象，并配有详细的文字介绍。该图册是地质学、地球化学、资源勘查、水文地质、地质工程、石油工程等专业学生在周口店野外实习期间良好的辅导教材，将为师生进行实习前的资料熟悉、实习期间的认真比对、实习后期的资料整理发挥重要作用。同时，书中编录的地质现象都具有典型意义，在普通地质学、地质学基础、构造地质学、地史学、地层古生物学等地质基础课程教学中也可以起到很好的补充作用。

　　该图册的出版，进一步完善了我校地质专业实习教材体系，给我校参加实践教学工作的师生带来极大的帮助，对周口店野外地质实验教学示范中心建设具有积极的促进作用。同时，也凝聚了"图册"课题组老师们对周口店野外地质实践教学基地建设的心血、对培养优秀地质人才的热切期盼。我想，阅读过此图册的教师和同学们都会对他们表示敬意。

中国地质大学（武汉）教授

欧阳建平

2010年7月16日

PREFACE

The Zhoukoudian field practice teaching is the most traditional and distinctive teaching content in China University of Geosciences. The Zhoukoudian area has unique geological phenomena and geographical landscape, a large number of different types of rocks and a wealth of geological features due to its unique geographical location and geological structure units. Typical strata of the North China since the Cambrian are well distributed and its geological evolution and development in time and space is highly representative. So Zhoukoudian is a rare field practice base for basic geology teaching in our country. In the early 1950s, Chi Jishang and Ma Xingyuan, academicians of the Chinese Academy of Sciences, and В.Н.Павлинов, former Soviet geologist, and other geologists of the older generation carried out field practice teaching in Zhoukoudian, which pioneered for Zhoukoudian practice base construction of Beijing College of Geology (now China University of Geosciences). In the past over 50 years, more than 40,000 students have participated in this field practice, and many of them became outstanding talents including national leaders, dozens of academicians, elite climbers of China, etc. Any graduate of our university who has participated in the Zhoukoudian field practice would place it firmly in mind and never forget the most memorable time of their lives. On the celebration of the 50th anniversary of the Zhoukoudian Practical Station in 2004, our schoolfellow and the State Premier Wen Jiabao who has practiced in Zhoukoudian for five times readily handwrote an inscription "cradle", recognizing the contribution of the Zhoukoudian Practical Station to geology in China.

The Zhoukoudian field teaching has many characteristics, such as strong practicality and strong correspondence between theoretical knowledge and field geological phenomena, which are very helpful for students to grasp geological field techniques and professional knowledge firmly. Therefore, since it was founded, the station has been highly valued by the university. Academicians of Chinese Academy of Sciences, Yang Zunyi, Wang Hongzhen, Zhao Pengda, Yin Hongfu, Yu Chongwen, Jin Zhenmin and so on have come to teach at Zhoukoudian successively, and university leaders come to the station to provide care and guidance every year. After many years of exploration and analysis by generations of geologists, a teaching method which emphasizes rigorous exploration, steady progress, repetitive practice and cognition has been formed and has fostered thinking and work styles of students to be hardworking and plain-living, cooperative and innovative.

In the 21st century, how to effectively guarantee the quality and continuously improve the level of field practice teaching to train students with basic field skills needed in modern geological work is the most important aspect of the Zhoukoudian field practice teaching of our university. To further enrich the Zhoukoudian geological field practice teaching resources for teachers and students with more typical, direct and sharp pictures, a team leaded by Professor Zhao Junming finally presented the "*Atlas of Classic Geological phenomena in the Zhoukoudian Field Practical Teaching Base* " " to us on the eve of the 60th anniversary of China University of Geosciences and the 58th anniversary of the Zhoukoudian practice station after three years of hard work. In accordance with the requirements of Zhoukoudian field practice teaching program, with colorful field pictures as the main body, accompanied by detailed text descriptions and explanations, the atlas has recorded every classic geological phenomenon in Zhoukoudian, including those in teaching routes; it mainly consists of five parts: Stratigraphy and Paleontology, Structural Geology, Rocks and Minerals, Quaternary Strata, Resources and Environment. The atlas is a good supplementary book for students majoring in geology, geochemistry, resource exploration, hydrogeology, geological engineering and petroleum engineering to do field practice in Zhoukoudian. It would play an important role in helping students and teachers to know data before practice, compare data during practice, and compile data after practice. Meanwhile, it is also a very good supplement for basic geological courses such as Physical Geology, Basic Geology, Structural Geology, Historical Geology, and Stratigraphic Paleontology as geological phenomena recorded in the atlas have typical significance.

The publication of the atlas further improves the teaching material system of geology practice of our university, provides tremendous help for our students and teachers who participate in practice teaching, and has a positive role in promoting the construction of Zhoukoudian geological field practice teaching demonstration center. The altas also embodies the endeavor of teachers in the team to construct Zhoukoudian geological field practice teaching base and the expectations of them to cultivate outstanding geological talents. I think all the students and teachers who have ever read the atlas would appreciate them.

Professor of China University of Geosciences (Wuhan)

OuYang Jianping
July 16, 2010

前言

位于北京市中心城区西南约45千米的周口店及周边地区，因其独特的大地构造属性，在漫长的地史演化过程中，由内外动力地质作用铸就了诸多地质现象。在一个相对不大的范围内(约350平方千米)，出露了类型齐全、数量众多，且具有典型意义的地质遗迹。该地区地学研究、地学教学资源得天独厚，加之区位优势和便利的交通条件，为我们提供了一个难得的野外实践教学场所。

最早研究包括周口店在内的北京西山地区地质的是美国地质学家庞派来(R.Pumpelly，1867年)，我国最早的地质专修班在此进行野外实习(1914年)。中国地质大学(原北京地质学院)于1954年在周口店创建野外实践教学基地，在50余年的历程中，广大师生员工艰苦奋斗克服重重困难潜心钻研，在教学、科研、人才培养等方面成绩斐然，为培养出数十名中国科学院院士及党和国家领导人、数百名著名地质学家和教育家、数以万计地学人才发挥了举足轻重的作用，周口店实习基地堪称地质科技工作者的摇篮。

为了更好地发挥周口店地区地质遗迹的教育功能，给"摇篮"增添光彩，我们在总结前人成果的基础上，利用影像具备直观性、可读性、视觉冲击力强的优势，借助影像技术手段，拍摄和收集了大量的经典地质现象图片，并配以简明扼要的文字阐述，以图文并茂的方式揭示地质现象特征及科学内涵。本书作为周口店野外实践教学图像辅导教材，有助于教师野外教学，有助于学生理解地质知识以及室内复习加深印象，巩固野外学习效果。

本图册由赵俊明、袁晏明负责体系设计与统编。图片摄制与收集由赵俊明在曾广策、章泽军、袁晏明、秦松贤、徐冉、李长安等实习教师带领下完成，文字描述分别由赵俊明(前言、第一部分、第六部分)，徐冉(第二部分)，袁晏明(第三部分)，秦松贤(第四部分)，李长安(第五部分)等撰写，英文翻译由袁爱华、王庆完成。

本图册编制过程中得到学校各方面的支持。国家基金委基础科学人才培养基金(J1030518)、国家级实践教学示范中心建设单位(教高函〔2007〕21号)、中国地质大学(武汉)设备处、教务处为本图册提供了经费资助。作者在提出本图册设想之初，中国地质大学(武汉)副校长欧阳建平教授给予了充分赞许和鼓励，更坚定了编制本书的信心。殷鸿福院士、欧阳建平副校长为本书作序。金振民、莫宣学院士为周口店实习基地的提词首次在本书中发表。叶俊林、王人镜、吴顺宝、谭应佳、单文琅、郭铁鹰、傅绍仁、曾广策、章泽军、赵温霞、杨坤光、桑隆康、闻立锋、张志、朱彩霞等许多教师为本图册的编制提出了宝贵建议及帮助，中国地质大学出版社刘桂涛以及大魏工作室为本书出版付出了辛勤的工作，在此一并表示衷心地感谢！

赵俊明

2010年6月20日

INTRODUCTION

The Zhoukoudian area, about 45 kilometers southwest to Beijing downtown, has varieties of geological phenomena formed by exogenic and endogenic geological processes in the long history of geological evolution with its unique tectonic attributes. In this relatively small area (about 350 square kilometers), a large number of different kinds of typical geological relics have been found. The unique and abundant geology & geosciences research and teaching resources, coupled with geographical advantages and convenient traffic conditions, provide us with a rare field practice teaching site.

The earliest study of Beijing Western Hill area, including Zhoukoudian, was conducted by the American geologist R. Pumpelly in 1876. The earliest students in our country majored in Geology also conducted field practice here in 1914. Since China University of Geosciences (former Beijing College of Geology) founded field practice teaching base in Zhoukoudian in 1954, the university has achieved remarkable success in teaching, research, and personnel training owing to the hard work of all teachers and students. The Zhoukoudian field practice teaching base has played an important role in educating tens of thousands of specialized people in geology, including hundreds of famous geologists and educators, dozens of academicians of the Chinese Academy of Sciences and top leaders of China in the past over 50 years. The Zhoukoudian field practice teaching base can be regarded as the cradle of geological scientific and technical workers.

In order to develop the educational function of Zhoukoudian geological relics more broadly and to give some sparkle to the "Cradle", we compile this photo-illustrated book after an overall review of existing achievement by taking and collecting amounts of classic photos of the geological phenomena using imaging techniques which feature visual and clear pictures with strong impact, matching with concise descriptions to reveal features and scientific connotations of geological phenomena. This book will function as image support materials to the teaching of Zhoukoudian field practice, which helps teachers to do field practice teaching more effectively, and students to understand geological knowledge better and thus enhance effects of field practice teaching through indoors review.

The atlas is designed and edited by Zhao Junming and Yuan Yanming. The pictures were photographed and collected by Zhao Junming under the guidance of Zeng Guangce, Zhang Zejun, Yuan Yanming, Qin Songxian, Xu Ran, Li Chang'an and other practice guidance teachers, while the text descriptions were written by Zhao Junming (Introduction, Part I, Part VI), Xu Ran (Part II), Yuan Yanming (Part III), Qin Songxian (Part IV) and Li Chang'an (Part V). English translation was conducted by Yuan Aihua and Wang Qing.

Compiling of the atlas was supported by our university in several ways. It was funded by the National Science Fund for Talent Training in Basic Science (Grant No.J1030518), National Demonstration Center for Practical Teaching (2007, No.21), and Facilities Management Office and Academic Administration of China University of Geosciences (Wuhan). When the writer proposed the idea of the atlas, vice-president of China University of Geosciences (Wuhan), Professor Ouyang Jianping, gave much appreciation and praise, which strengthened our confidence in writing the atlas. We thank Academician Yin Hongfu and vice-president Ouyang Jianping for writing prefaces for the atlas. The inscriptions written by academicians Jin Zhenmin and Mo Xuanxue for Zhoukoudian field practice teaching base are published in this atlas for the first time. Ye Junlin, Wang Renjing, Wu Shunbao, Tan Yingjia, Shan WenLang, Guo Tieying, Fu Shaoren, Zeng Guangce, Zhang Zejun, Zhao Wenxia, Yang Kunguang, Sang Longkang, Wen Lifeng, Zhang Zhi, Zhu Caixia, and many other teachers contributed to the atlas with comments and suggestions are greatly appreciated. We sincerely thank Liu Guitao from China University of Geosciences Press and Dawei Workroom for their hard work in the press of the book.

Zhao Junming
June 20, 2010

目录 / CONTENT

1 实习基地概况
Introduction of the Zhoukoudian Field Practical Teaching Base /1

1.1 周口店野外教学实习基地区域位置
The location of the Field Practical Teaching Base /2

1.2 周口店野外教学实习基地及学习条件
Education facilities in the Zhoukoudian Field Practical Teaching Base /7

2 地层、古生物
Strata and Paleontology /11

2.1 地层
Strata /12

2.2 古生物
Paleontology /34

3 矿物、岩石
Minerals and Rocks /39

3.1 矿物
Minerals /40

3.2 岩石
Rocks /48

4 构造地质
Structure Geology /67

4.1 劈理
Cleavages /69

4.2 节理
Joints /73

4.3 线理
Lineations /77

4.4 褶皱
Folds /82

4.5 断层
Faults /97

5 第四纪(含上新世)地层及地貌
Quaternary (including the Pliocene) and Geomorphology /115

5.1 第四纪(含上新世)地层
Quaternary (including the Pliocene) /116

5.2 地貌
Geomorphology /124

5.3 主要不同成因的沉积物
Major accumulations of different genesis /134

5.4 古人类与古动物
Palaeo-human and Palaeo animal /139

6 区域资源与环境保护
Regional Resources and Environmental Protection /147

6.1 矿产资源
Mineral resources /148

6.2 旅游资源
Tourism resources /153

6.3 水资源及环境保护
Water resources and environmental protection /159

1 实习基地概况

Introduction of the Zhoukoudian Field Practical Teaching Base

中国地质大学周口店实习区,位于北北东向太行山山脉、近东西向燕山山脉和华北平原接壤地带,北至大石河,南至拒马河,东至房山,西至十渡。实习范围内大多为中、低山区,一般为 50～200 米,地势西北高、东南低,北部最高山峰是上寺岭(猫耳山),海拔 1 307 米;东南部接壤华北平原。区内河流主要有拒马河、大石河、周口店河、黄山店河等。

The Zhoukoudian Field Practice Area of China University of Geosciences is located in the border area between NNE trending Taihang Mountain Range, near EW trending Yanshan Mountain Range and the North China Plain, north to Dashi River, south to Juma River, east to Fangshan Mountain and west to Shidu, high in the northwest and low in the southeast, with North China Plain on the southeast. Low and middle mountains from 50~200m occupy most of the practice area. The highest peak is Shangsiling (Maoer Mountain) in the north, at 1 037m. The main rivers are Juma River, Dashi River, Zhoukoudian River, and Huangshandian River.

1.1 周口店野外教学实习基地区域位置
The location of the Field Practical Teaching Base

中国地质大学周口店实践教学基地设在距北京市中心城区西南约 45 千米的房山区周口店镇,地理坐标为北纬 39°41′,东经 115°51′,实习区域铁路、公路交通十分便利。

The Zhoukoudian Field Practical Teaching Base of China University of Geosciences is located in Zhoukoudian Town, Fangshan District, about 45km SW of downtown Beijing, at 39°41′N and 115°51′E. It can be directly reached by bus or train.

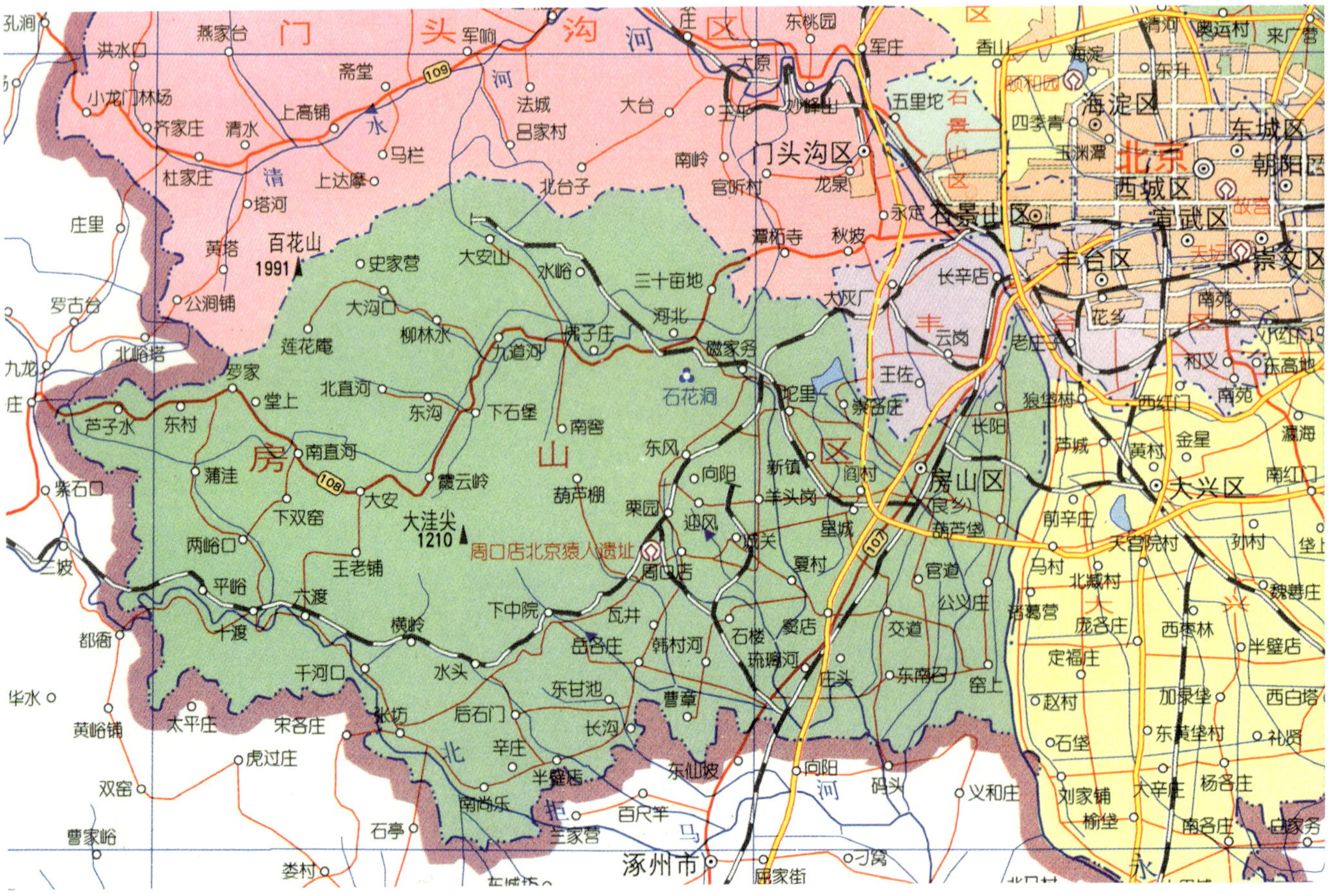

周口店实习区交通位置图(据《中国分省地图集》星球地图出版社 2006 年第 3 版)
Location and transportation of the Zhoukoudian Field Practical Teaching Base (Refers To "Atlas of China by Province", Star Map Press, 3rd Edition, 2006)

周口店镇、太平山、房山岩体地形地貌鸟瞰图
Topography of Zhoukoudian Town, Taiping Mountain and Fangshan Pluton

太平山北麓主要实习区地形地貌
Topography of major practice area covering the northern piedmont of the Taiping Mountain

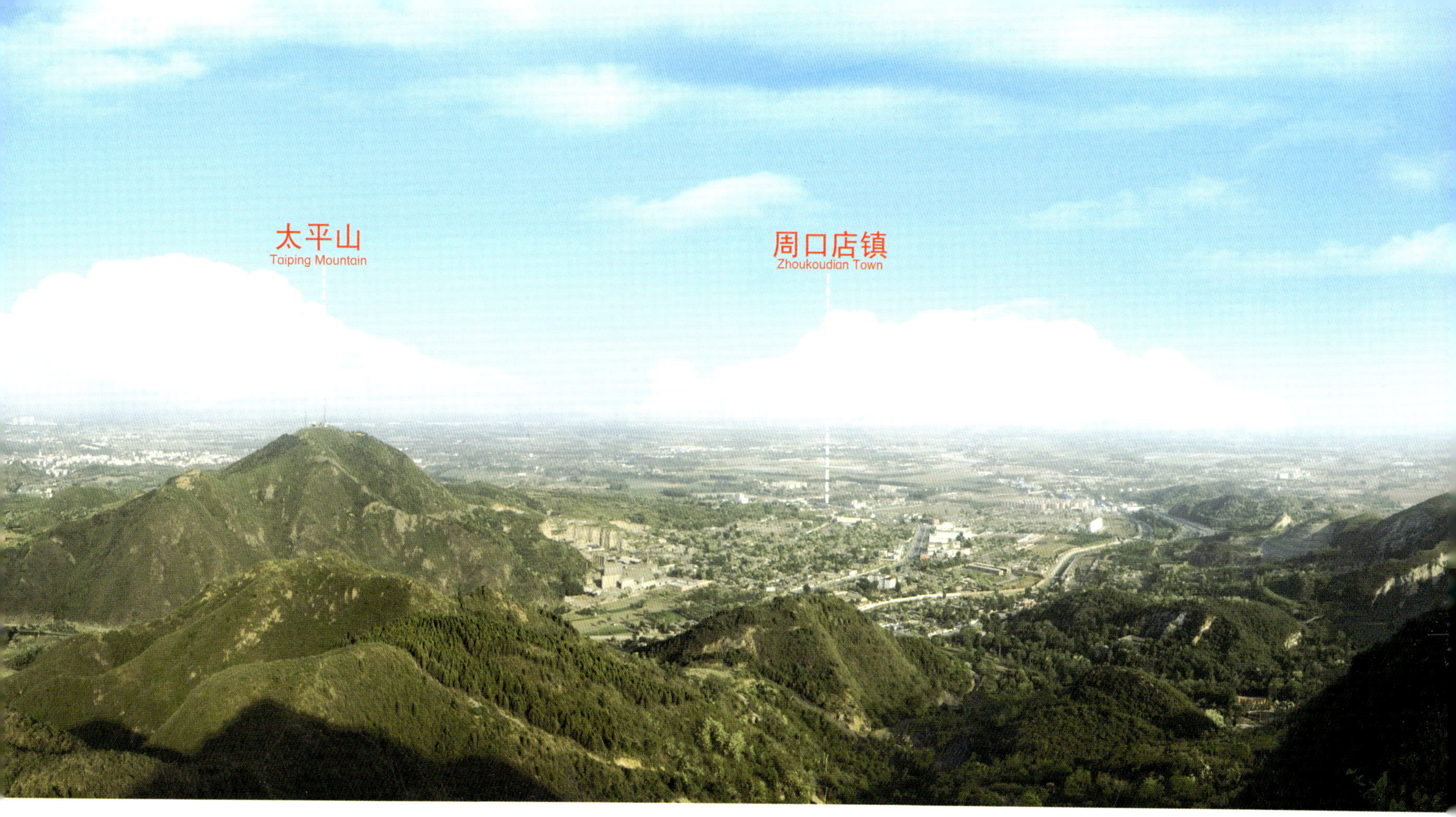
太平山
Taiping Mountain
周口店镇
Zhoukoudian Town

太平山
Taiping Mountain
一条龙
Yitiaolong
周口河
Zhoukou River

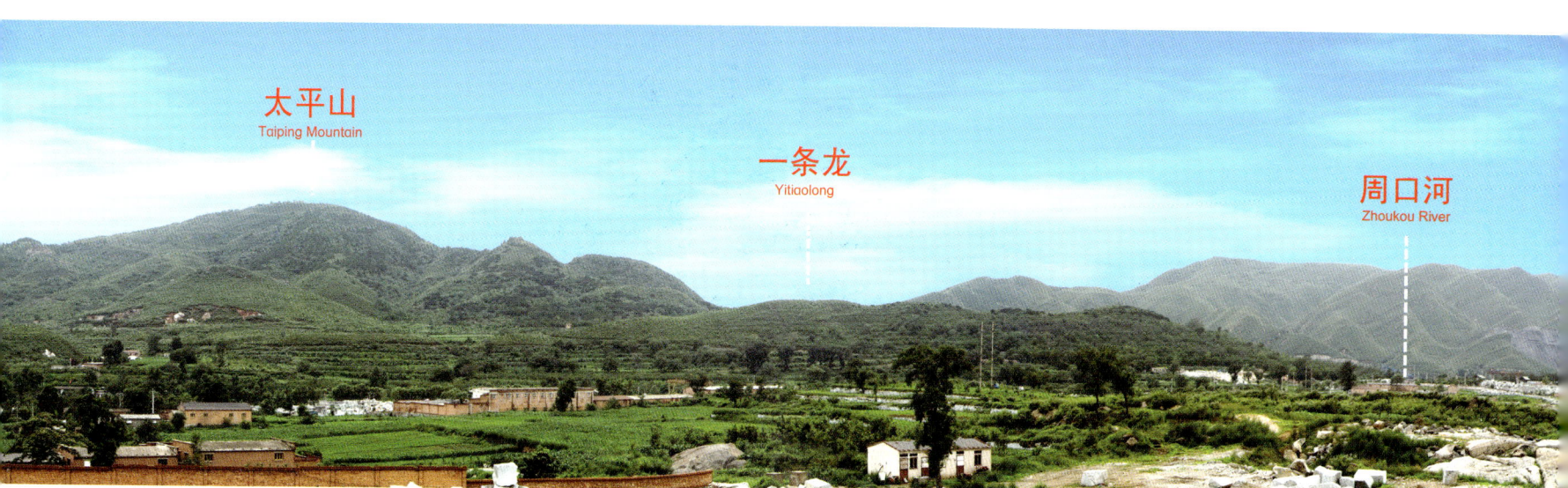

周口店 1∶5 万地质简图
Simplified Geological Map (1:50 000) of Zhoukoudian

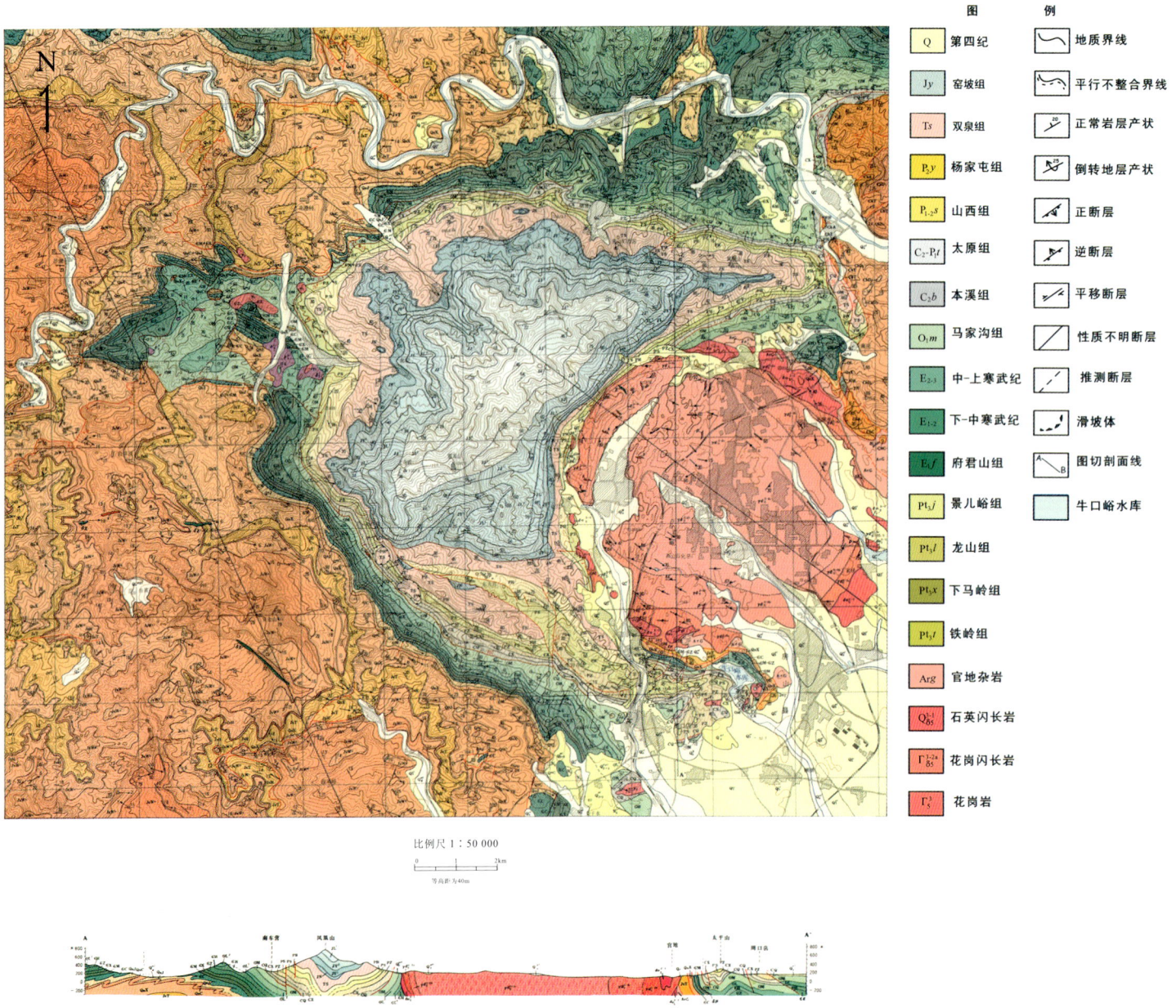

周口店所属的北京西山地区处在华北板块中部，位于北北东向太行山隆起东北端，东邻华北平原，北为近东西向的燕山山脉，在构造上处于一个十分独特的位置，这两个方向控制了区内多期构造的运动图像。

Zhoukoudian is situated in the Xishan area of Beijing, middle of the North China Plate. It's at the northeastern uplift zone of the NNE trending Taihang Mountain Range, with North China Plain on the east, and EW trending Yanshan mountain Range on the north. So its geological position and structure is very unique and the two directions are in control of the multistage structural activities.

1.2 周口店野外教学实习基地及学习条件
Education facilities in the Zhoukoudian Field Practical Teaching Base

周口店实习基地院内占地面积 15 亩,建筑面积 2 500 平方米,具备良好的野外实习学习条件和生活环境,同批次可接纳 500 余名师生开展野外教学实习。为了与野外路线教学相衔接,实习基地内建立了地质陈列室、图书资料阅览室、岩矿鉴定室、信息技术处理室(微机室)、地质展景等配套教学设施。

The whole Zhoukoudian Field Practical Station occupies 15 Chinese mu, while the floorage occupies 2 500 square meters. The station provides favorable field practice learning conditions and comfortable living environment for students and teachers, and can accommodate more than 500 people at a time. In order to link up with the field practice teaching, geology exhibition room, reading room, rock and mineral identification room, information processing room (computer room), geological gardens and other training sites and related facilities are provided in the station too.

周口店实习站大门(原)
Gate of the Zhoukoudian Practical Station (old)

周口店实习站大门(现)
Gate of the Zhoukoudian Practical Station (new)

实习站宿舍
Dormitories of the station

图书资料阅览室
Reading Room

岩矿鉴定室（曾广策提供）
Rock and Mineral Identification Room
(Provided by Zeng Guangce)

微机室（曾广策提供）
Computer Room
(Provided by Zeng Guangce)

地质陈列室
Geology Exhibition Room

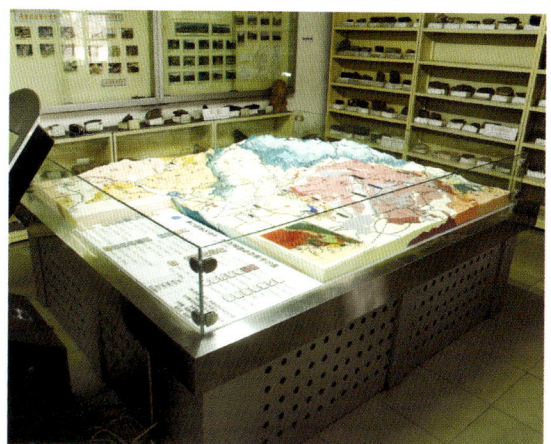
地质沙盘
Geological Sand Table

地质展景园地
Geological Garden

教师集体备课
Group teaching preparation

教师在实习基地给学生授课
Teaching in the Station

教师在野外指导学生观察地质现象
The teachers are guiding the students to observe the geological phenomenon in the field.

教师给学生辅导答疑
The teachers are tutoring the students

学生在实习站内整理资料
The students are collating data in the practical station

大学生在实习站内开展体育活动（袁晏明摄）
The students are doing sports in the practical station (The photo is taken by Yuan Yanming)

实习基地举行建站50周年庆祝活动（赵温霞提供）
The people are celebrating the 50th anniversary of establishment of the practical station(The photo is provided by Zhao Wenxia)

2 地层、古生物
Strata and Paleontology

周口店地区地层属于华北型，经历了基底阶段，似盖层、盖层发育阶段，出露较为齐全，涵盖了太古宇、中元古界、新元古界、古生界、中生界和新生界。地层接触关系多样，沉积环境各异，是进行地史学、地层学和沉积学野外实践教学的良好场所。

北京周口店及其邻近地区产许多重要古生物化石，除世界著名北京人头骨化石以外，从元古代至第四纪地层中皆有化石产出，如叠层石、三叶虫、珊瑚、腕足类；石炭纪至二叠纪的植物化石，第四纪的鱼化石等。这些化石的发现为地层时代的确定和对比提供了有力证据。（化石图片均摄于周口店实习站陈列室）

Strata well exposed in the Zhoukoudian area belong to the North China type in stratigraphy, which evolved with developments of the basement, paraucover and covers including the Archean, Mesoproterozoic, Neoproterozoic, Paleozoic, Mesozoic and Cenozoic. With various stratigraphic contacts and sedimentary paleoenvironment, the Zhoukoudian area is an excellent place for field practical teaching in Historical Geology, Stratigraphy and Sedimentology.

There are many important fossils in Zhoukoudian of Beijing and its adjacent areas. In addition to the famous Peking Man, the Proterozoic to Quaternary yield different kinds of fossils as stromatolites, trilobites, corals, brachiopods, Carboniferous-Permian fossil plants and Quaternary ichthyolites, which provide convincing evidences for stratigraphic division and comparison.(All photos in this part are taken in the geology showroom of the Station)

2.1 地层 Strata

年代地层 界	系	统	岩石地层 群	组	段	代号	柱状图	厚度	岩性描述及化石	矿产
新生界	第四系(Q)	全新世				Q_4		4-20	残存堆积亚砂土层，洪冲积砂砾层、土壤层	
		更新世		周口店组		Q_p		4-16		
				太平山组		Q_1T			坡积红色角砾砂土层，洞穴内沉积层、风化杂色粘土层	
	第三系(N)	上新世				N_2		34.7	喜山运动 砂砾层、粘土层	
									角度不整合	
上古生界	三叠系(T)			双泉组		Ts		181	上部灰绿色含泥砾板岩；中部灰绿、紫灰色凝灰质板岩；下部灰绿色中粗粒变质砂岩。下部产Gigantopteris shangqianensis（双泉大羽羊齿）	
	二叠系(P)	上统		红庙岭组		$P_{2-3}h$		>200	肉红色、灰白色变质石英粗砂岩，变质细砂岩，粉砂质板岩 灰色厚层变质中粗粒岩屑砂岩，含岩屑砂岩，夹碳质板岩，底部多为灰色厚层变质复成分角砾岩	
		中统		杨家屯组		P_2y		120	褐灰色中厚层变质中粗粒岩屑砂岩，黑色炭质板岩。上部深灰色中厚层变质中细粒变质砂屑砂岩，上部为黑色炭质板岩、粉砂质板岩夹煤层。含Lobatannularia sinensis（中华瓣轮叶），Sphenophyllum thonii（汤氏楔叶）	煤
		下统		山西组		$P_{1-2}s$		90	灰色、褐灰色中厚层变质细粒石英砂岩夹灰黑色板岩；上部主要为灰黑色、褐灰色薄层粉砂质、粉砂岩，并夹有薄煤层。产Neuropteris ovata（卵脉羊齿），N.plicata（褶脉羊齿）	红柱石
				太原组		C_3-P_1t		64		
	石炭系(C)	上统		本溪组		C_2b		54	上部为压力影板岩，红柱石角岩及透镜状生物碎屑灰岩，中部为杂色粉砂质板岩及变质粉砂岩，下部为硬绿泥石角岩及红柱石角岩。局部为褐铁矿、铝土矿风化壳底砾岩	粘土矿
下古生界	奥陶系(O)	下统		马家沟组		O_1m		300	平行不整合（怀远运动） 青灰色厚层结晶灰岩、纹带状灰岩夹少量白云质灰岩，产 Armenoceras sp.（阿门角石）	水泥原料
				亮甲山组		O_1l		70	灰色中-厚层结晶白云岩，夹2~3层灰色竹溶角砾岩，含少量燧石团块	
				冶里组		O_1y		67	青灰色纹带状灰岩夹少量灰色豹皮状白云质灰岩及黄色板岩，底部普遍存在钙质千枚岩	
	寒武系(∈)	上统		黄院组		$∈_3h$		123	灰黄色薄层泥质条带灰岩夹少量薄层鲕粒灰岩，量竹叶状灰岩 灰绿色千枚状板岩、粉砂质板岩夹中厚层鲕状灰岩和结晶灰岩互层。产Damesella sp.（德氏虫）	
		中统		张夏组		$∈_2zh$		36	灰绿色页岩夹薄中层灰岩	
				徐庄组		$∈_2x$		41	灰色灰黄色页岩夹白云质灰岩透镜体	
				馒头组		$∈_{1-2}m$		46		
		下统		府君山组		$∈_1f$		45	上部为灰色中厚层纹带灰岩，豹皮灰岩，下部为豹皮灰岩夹白云质灰岩 平行不整合（蓟县运动）	水泥原料
新元古界	青白口系(Qb)	上统	青白口群	景儿峪组		Pt_3j		55	上部为灰绿色钙质千枚岩，下部为灰白色中薄层大理岩	板材
				龙山组		Pt_3l		>20	上部为浅灰色斑点状板岩，下部为灰色中厚层变质石英砂岩	
		下统		下马岭组	上 中 下	Pt_3x		170	上部灰色粉砂质千枚状板岩，黑色碳质千枚状板岩，中部褐色粉砂质板岩夹薄层变质细砂岩，发育小型交错层理。下部为褐色千枚状板岩，含磁铁矿板岩，底部具有风化壳	
	蓟县系(Jx)	上统	蓟县群	铁岭组	上 中 下	Pt_2t		186	平行不整合（芹峪运动） 上部灰色中厚层结晶白云岩，叠层石条带，中部含燧石条带，下部为灰白色厚层结晶白云岩，交错层理发育	
				洪水庄组		Pt_2h		38	灰褐色锰质板岩，顶部和底部夹含锰白云岩及白岩透镜体	
		下统		雾迷山组		Pt_2w		>500		
中元古界	长城系(Ch)	上统	长城群	大红峪组		Pt_2d		355	灰色、浅灰色中薄层结晶白云岩，夹灰质白云岩，夹较多燧石条带或燧石团块 平行不整合 砖红、灰红色长石石英砂岩夹灰白色石英岩	
				团山子组		Pt_2tu		91	灰白、深灰色含砂白云岩与石英岩，千枚岩互层	
		下统		串岭沟组		Pt_2c		106	灰黑色砂质板岩和变质细砂岩互层	
				常州沟组		Pt_2ch		30	灰白色中厚层石英砂岩，仅在南大寨及东流水一带出露，交错层理发育	
太古宇				官地杂岩		Arg		>200	断层接触 条带状混合岩，黑云斜长片麻岩、斜长角闪岩、角闪岩	

周口店实习区综合地层柱状图
（引自周口店实践教学示范中心）

An integrated stratigraphic histogram of the Zhoukoudian practice area (from Zhoukoudian Demonstration Center for Practical Teaching)

0 50 100m

2.1.1 太古界
Archean

周口店实习区内的太古界变质杂岩(官地杂岩,即称原关坨杂岩)零星分布于房山岩体边部(官地村、山顶庙北侧冲沟旁),面积约0.37平方千米,它们与上覆不同时代的地层呈剥离断层接触。官地杂岩的主要岩性由黑云角闪斜长片麻岩、混合片麻岩、斜长角闪岩、黑云母角闪石变粒岩等组成,因遭受了强烈的动力变质作用,普遍发生糜棱岩化。

The Archean metamorphic complex (i.e. the Guandi Complex) are scattered at the edge of the Fangshan Pluton (in the Guandi Village and gullies to the north of Shandingmiao) with an area of about 0.37 km². The complex are underlain by detachment faults. In lithology, the Guandi Complex consist of biotite-hornblende-plagioclase gneisses, migmatites, amphibolites and biotite-hornblende leptites. These rocks experienced strong dynamic metamorphism with common mylonitization.

官地杂岩
The Guandi Complex

2.1.2 中、新元古界
Mesoproterozoic and Neoproterozoic

黄山店雾迷山组剖面
Outcrop of the Wumishan Formation in the Huangshandian Section

黄山店剖面出露雾迷山组第四段厚层白云岩，风暴沉积发育，多见破碎硅质条带和同沉积断层。（摄于黄山店）
The exposed is the fourth Member of the Wumishan Formation in the Huangshandian Section that is presented as thick dolomite. The well-developed tempestite deposits present common siliceous bands, which have been offset by synsedimentary faults. (Photographed in Huangshandian)

同沉积断层
Synsedimentary faults

黄山店剖面雾迷山组第四段中发育的同沉积断层(箭头A),该断层是在沉积物尚处于塑性-半塑性状态下时形成,穿切了硅质条带(箭头B)在下方不远处即尖灭消失。(摄于黄山店)

The synsedimentary fault, which occurs in the fourth Member of the Wumishan Formation (indicated by the arrow A), was formed as sediments remained in plastic to semi-plastic state. The fault offsets siliceous bands (indicated by the arrow B), below which it disappears.

风暴沉积
Tempestite deposits

雾迷山组中硅质条带在半塑性状态下受到风暴流搅动后再次沉积,其排列往往呈倒"小"字型等杂乱无章排列。(摄于黄山店)

In the Wumishan Formation, siliceous bands that are disturbed by storm surges in semi-plastic state were re-deposited with disordered arrangement.(Photographed in Huangshandian)

压溶缝合线
Stylolites by pressure solution

成岩过程中,由于上覆岩层静压力,造成岩层面之间界发生不均匀溶解而成。常见于碳酸盐岩中。(摄于拒马河三渡半雾迷山组)

Stylolites, which usually appear in carbonatites are formed by uneven dissolution occurring between rock surfaces by static pressure from overlying layers during diagenesis. (Photographed in Sanduban, the middle between Sandu and Sidu,Wumishan Formation)

洪水庄组与雾迷山组界线
Contact between the Hongshuizhuang and Wumishan Formations

上部为洪水庄组（Pt₂h）含锰板岩（分布在黄山店、八角寨一带。岩性以灰黑色含锰质板岩为主），下部为雾迷山组（Pt₂w）第四段白云岩，两者整合接触。（摄于八角寨）

The upper strata consisting of grey black manganiferous slates are of Hongshuizhuang Formation (Pt$_2$h), which is distributed in Bajiaozhai and Huangshandian. The lower presented as dolomite are of the fourth Member of the Wumishan Formation (Pt$_2$w). The contact between the two formations above is conformable. (Photographed in Bajiaozhai)

铁岭组底部内碎屑
Intraclast limestones at the bottom of the Tieling Formation

先期沉积物，在半固结状态下被后期风暴等强动力水流打碎、搅拌后在原地重新沉积而成。（摄于八角寨）

Intraclasts are formed by in situ re-deposition of original sediments that are disturbed by strong currents such as storm surges in semi-consolidated state. (Photographed in Bajiaozhai)

铁岭组顶部叠层石
Stromatolites at the top of the Tieling Formation

由菌藻类微生物分泌、粘结碳酸盐颗粒、碎屑物等形成。叠层石中每一个纹层均指示菌藻类微生物的一个生长周期。（摄于八角寨）

Stromatolites were formed by cementation of carbonatite grains and other detritus by secretions of microorganisms as fungi and algae. (Photographed in Bajiaozhai)

铁岭组与洪水庄组界线
Contact between the Tieling and Hongshuizhuang Formations

铁岭组（Pt$_2$t）：主要分布在黄山店、八角寨一带，一条龙、周家坡等地亦有出露。岩性为灰色含锰质白云岩夹硅质条带。底部发育交错层理，顶部发育叠层石。

上部为铁岭组（Pt$_2$t）白云岩，下部为洪水庄组（Pt$_2$h）含锰板岩，两者整合接触。（摄于八角寨）

The Tieling Formation, which is grey manganiferous dolomites interlayered by siliceous bandings in lithology, is mainly distributed in Huangshandian and Bajiaozhai and also exposed in Yitiaolong and Zhoujiapo. Cross beddings occur at the bottom of the formation while stromatolites at the top.

The upper are dolomites of the Tieling Formation (Pt$_2$t), while the lower are manganiferous slates of the Hongshuizhuang Formation (Pt$_2$h). The contact between them is conformable. （Photographed in Bajiaozhai）

铁岭组鱼骨状交错层理
Herringbone cross beddings in the Tieling Formation

由双向水流牵引沉积物运动形成。通常见于碳酸盐岩及碎屑岩中。（摄于八角寨）
Herringbone cross beddings formed by movement of sediment grains in force of bidirectional paleocurrents. (Photographed in Bajiaozhai)

铁岭组底部发育在石英砂岩中的单向交错层理
Cross beddings in quartz sandstone at the bottom of the Tieling Formation

在流体推动牵引作用下，沉积物颗粒滚动而形成。常见于碳酸盐岩及碎屑岩中。（摄于八角寨）
Cross beddings formed by dragging and rolling of sediment grains in force of paleocurrents. This structure is often found in carbonatite and clastic rocks. (Photographed in Bajiaozhai)

八角寨下马岭组与铁岭组平行不整合剖面及古风化壳

Parallel unconformity and paleo-weathered crust between the Xiamaling Formation and the Tieling Formation in Bajiaozhai

图 A：剖面全貌，虚线框位置为下马岭组与铁岭组界线，即平行不整合面及古风化壳所在。该平行不整合为芹峪抬升所致。

图 B：图 A 中虚线框部分放大，下马岭底部古风化壳。

图 C：图 B 的局部放大。

Fig. A: Panorama of the section. Dashed rectangles indicate the contact between the Xiamaling Formation and the Tieling Formation, i.e. the position of the parallel unconformity contact and paleo-weathered crust, which results of the Qinyu Movement.

Fig. B: Magnification of the dashed rectangle in figure A shows the paleo-weathered crust at the bottom of the Xiamaling Formation.

Fig. C: Magnification of the dashed rectangle in figure B.

下马岭组底部千枚状板岩中褐铁矿化的磁铁矿
Magnetites with ferritization in phyllitic slates at the bottom of the Xiamaling Formation

下马岭组（Pt_3x）：分布在黄院、拴马庄、长流水以及一条龙、山顶庙和房山一带。岩性以千枚状板岩及粉砂质板岩为主。底部含磁铁矿。（摄于八角寨）

The Xiamaling Formation (Pt_3x) is distributed in Huangyuan, Shuanmazhuang, Changliushui and Yitiaolong-Shandingmiao-Fangshan, which is represented by phyllitic slates and silty slates in lithology. Magnetites are common at the bottom of this formation. (Photographed in Bajiaozhai)

龙山组底部变质石英砂岩中发育的交错层理
Cross beddings in metamorphic quartz sandstones at the bottom of the Longshan Formation

龙山组（Pt_3l）：分布在黄院、拴马庄、长流水以及一条龙、山顶庙和房山一带。底部灰色–褐灰色厚层变质中粗粒石英砂岩，发育交错层理；上部为浅灰色千枚状板岩。（摄于黄院）

The Longshan Formation ((Pt_3l), which occurs in Huangyuan, Shuanmazhuang, Changliushui and Yitiaolong-Shandingmiao-Fangshan, are grey-brownish and grey, thick bedded, medium-coarse grained metamorphic quartz sandstones at the bottom and grayish phyllitic slates at the top of the formation. (Photographed in Huangyuan)

景儿峪组灰黑色薄大理岩
Grayish black thin bedded marbles in the Jingeryu Formation

景儿峪组（Pt_3j）：分布在黄院、拴马庄、长流水以及一条龙、山顶庙和房山一带。下部为白色薄–中层大理岩夹灰黑色薄层大理岩；上部为灰色、灰黄色钙质板岩。（摄于黄院）

The Jingeryu Formation (Pt_3j), which occurs in Huangyuan, Shuanmazhuang, Changliushui and Yitiaolong-Shandingmiao-Fangshan, consists of white thin-medium bedded marble interlayered with grayish black thin bedded marble in the lower, and of grey and isabelline calcareous slates in the upper. (Photographed in Huangyuan)

黄院东山梁新元古界-下古生界剖面
A panorama of the Early Paleozoic section at the eastern ridge of Huangyuan

2.1.3 下古生界
Lower Paleozoic

周口店地区下古生界只发育寒武系和下奥陶统,其分布较广,周口店、黄院、南窑及磁家务一带均有出露,黄院、长流水等地发育较好。

Lower Paleozoic in the Zhoukoudian area consists only of the Cambrian and Lower Ordovician, which are widely distributed in the Zhoukoudian Town, Huangyuan, Nanyao and Cijiawu with best exposure in Huangyuan and Changliushui.

府君山组 ($\epsilon_1 f$) 　馒头—毛庄组 ($\epsilon_{1+2} m$) 　徐庄组 ($\epsilon_2 x$) 　张夏组 ($\epsilon_2 zh$) 　黄院组 ($\epsilon_3 h$)

黄院东山梁下古生界剖面全景
Neoproterozoic-Lower Paleozoic section at the eastern ridge of Huangyuan

冶里组（O_1y）

亮甲山组（O_1l）

马家沟组（O_1m）

中、下寒武统府君山组、馒头–毛庄组和徐庄组剖面

The Lower-Middle Cambrian Section consisting of the Fujunshan, Mantou-Maozhuang and Xuzhuang Formations

上部为中寒武统徐庄组灰绿色千枚状板岩,中部为下–中寒武统馒头–毛庄组灰黄色千枚状板岩,下部为下寒武统府君山组泥质条带灰岩。三者之间均为整合接触关系。(摄于黄院东山梁)

徐庄组($Ɛ_2x$):灰色、灰绿色页岩、千枚状板岩夹薄层灰岩。

馒头–毛庄组($Ɛ_{1+2}m$):周口店地区馒头组及毛庄组界线不易划分,故而合之。灰色、灰黄色页岩,千枚状板岩夹白云岩透镜体。

府君山组($Ɛ_1f$):下部豹皮灰岩,上部泥质条带灰岩。

The section consists in the upper part of grayish green phyllitic slates of the Xuzhuang Formation of Middle Cambrian, in the middle of isabelline phyllitic slates of the Mantou-Maozhuang Formation of Lower-Middle Cambrian, and in the lower of banded pelitic limestones of the Fujunshan Formation of Lower Cambrian. Conformable contacts occur between these formations. (Photographed at the eastern ridge of Huangyuan)

$Ɛ_2x$: The Xuzhuang Formation ($Ɛ_2x$) consists of grey and grayish green shale and phyllitic slates interlayered with thin-bedded limestones.

$Ɛ_{1+2}m$: The contact between the Mantou Formation and the Maozhuang Formation is not well delimited in the Zhoukoudian area so that the two formations are combined as Mantou-Maozhuang Formation ($Ɛ_{1+2}m$). This formation consists of grey and grayish green shale and phyllitic slates interlayered with dolomite lens.

$Ɛ_1f$: Lower part of the Fujunshan Formation ($Ɛ_1f$) consists of leopard-skin limestones, while the upper part of banded pelitic limestones.

中寒武统张夏组鲕粒灰岩

Oolitic limestones in the Zhangxia Formation of Middle Cambrian

张夏组（Є_2zh）：鲕粒灰岩与千枚状板岩互层。鲕粒一般形成于浅滩，在高能流体反复冲刷下，沉积物颗粒反复滚动形成同心纹状微球粒。（摄于黄院东山梁）

The Zhangxia Formation (Є_2zh) is represented in lithology by interlayered oolitic limestones and phyllitic slates. Oolith appearing as concentric shaped commonly occurs in shoal environments, where sediment grains remain in lasting rolling by repeating wash of strong currents. (Photographed at the eastern ridge of Huangyuan)

上寒武统黄院组泥质条带灰岩（摄于黄院东山梁）

Banded pelitic limestones in the Huangyuan Formation of Upper Cambrian (Photographed at the eastern ridge of Huangyuan)

黄院组（Є_3h）：相当于山东张夏地区的崮山组、长山组和凤山组。下部为灰-灰黄色泥质条带岩夹少量薄层鲕粒灰岩。上部为灰色薄-中层纹带状灰岩。

The Huangyuan Formation (Є_3h) is equivalent to the Gushan, Changshan and Fengshan Formations in the Zhangxia area of Shandong Province. This formation in the lower part consists of grey-isabelline banded pelitic limestones interlayered with a few beds of oolitic limestones, and in the upper part of grey thin-medium bedded lamella limestones.

下奥陶统冶里组（O_1y）
The Yeli Formation of Lower Ordovician

冶里组（O_1y）：浅灰色-青灰色中厚层纹带状灰岩夹少量灰色豹皮状白云质灰岩及灰黄色板岩，产角石：*Piloceras* sp.（枕角石），*Cameroceras* sp.（房角石），腹足类：*Ophileta* sp.（蛇卷螺），古杯类：*Archaeocyathus* sp.（原古杯），厚 67 米。区域上本组底部是一层灰色钙质板岩（在某些区段因变质程度差异可为钙质千枚岩），而与下伏黄院组区分。

The Yeli Formation (O_1y), with a thickness of 67m, consists of grayish to caesious medium-thick bedded lamella limestones interlayered with grey leopard dolomitic limestones and isabelline slates. The hornstones *Piloceras* sp. and *Cameroceras* sp., the gastropods *Ophileta* sp. and the archaeocyathids *Archaeocyathus* sp. are yielded in this formation. Regionally, this formation is differentiated from underlying Huangyuan Formation by grey calcareous slates (calcareous phyllite in some localities due to the differential metamorphic grades) at the bottom of the formation.

怀远运动形成的平行不整合
Parallel unconformity by the Huaiyuan Movement

怀远运动：晚奥陶世地壳抬升，发生大规模海退，使华北板块成为持久的古陆剥蚀区，直到晚石炭世才沉降到海面以下，重新接受沉积。（摄于太平山南坡）

马家沟组（O_1m）：青灰色厚层结晶灰岩、纹带状灰岩夹少量白云质灰岩，局部地段夹灰褐色钙质板岩，产角石。与下伏亮甲山组整合接触。

亮甲山组（O_1l）：灰色中-厚层结晶白云岩，夹2～3层灰色膏溶角砾岩，含少量燧石团块。

The Huaiyuan Movement, which occurred in Late Ordovician, initiated regional crust uplift and large-scale ocean regression making long-term regional denudation that lasted until Late Carboniferous, when the North China Block subsided below sea level with acceptance of deposition. (Photographed in the southern slope of the Taiping Mountain)

The Majiagou Formation (O_1m), underlain the unconformable contact, consists of recrystallized caesious thick bedded limestones, lamella limestones interlayered with dolomitic limestones and brownish grey calcareous slates locally. Hornstones are yielded. This formation that is conformable with the underlying Liangjiashan Formation.

The Liangjiashan Formation (O_1l) is represented in lithology by recrystallized grey medium-thick bedded limestones interlayered with a few nodular cherts and two or three beds of grey gypsum breccia.

2.1.4 上古生界
Upper Paleozoic

周口店地区上古生界地层缺失泥盆系及下石炭统。上石炭统—侏罗系主要分布在上寺岭-凤凰山的南、北坡和黄院、升平山及太平山一带。岩层普遍遭受过轻度区域变质作用，局部受到岩浆侵入作用的影响。

The Devonian and Lower Carboniferous are absent in the Upper Paleozoic of the Zhoukoudian area. The Upper Carboniferous to Jurassic are mainly distributed in the northern and southern slopes of the Shangsiling-Fenghuang Mountain, Huangyuan, Shengping Mountain and Taiping Mountain, which commonly experienced low-grade regional metamorphism, and locally influenced by magmatic intrusion.

煤炭沟-太平山南坡地形地貌图
Topography of the southern slope of the Taiping Mountain

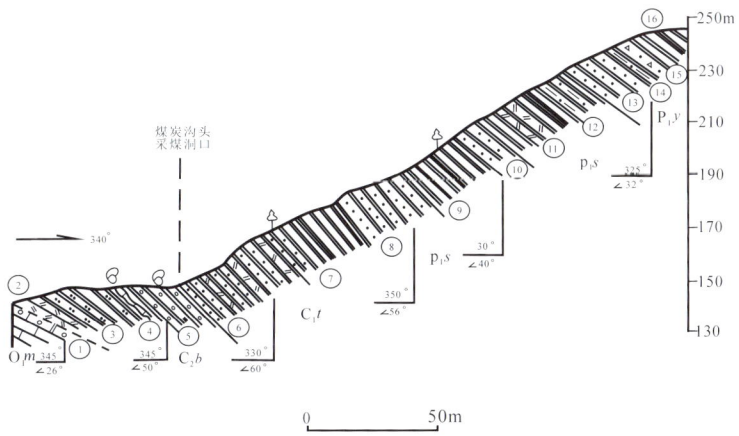

煤炭沟-太平山南坡石炭系-二叠系实测剖面图
Cross section of the Carboniferous-Permian across the southern slope of the Taiping Mountain

①砾状白云质灰岩；②硬绿泥石角岩及红柱石角岩；③杂色粉砂岩、砂质板岩；④凸镜体状泥质灰岩；⑤含黄铁矿黑色板岩；⑥变质杂砂岩与红柱石角岩互层；⑦杂色板岩夹炭质板岩、煤层；⑧变质中-粗粒杂砂岩；⑨黑色板岩夹煤层；⑩变质砂岩、粉砂岩；⑪黑色板岩夹煤线；⑫灰色砂质板岩；⑬变粗粒杂砂岩；⑭砂质板岩；⑮变质含砾杂砂岩；⑯黑色板岩。(引自《北京周口店地质及地质教学实习指导书》图2—10)

① Rudaceous dolomitic limestone； ② Chloritoid hornfels and andalusite hornfels； ③ Mottled siltstone and arenaceous slate； ④ Lentoid argillaceous limestone； ⑤ Black slate bearing pyrite； ⑥ Interlayered metagreywacker and andalusite hornfels； ⑦ Mottled slate interlayered with carbonaceous slate and coal； ⑧ Medium-coarse grained metagreywacker； ⑨ Black slate with coal beds； ⑩ Metasandstone and metasiltstone； ⑪ Black slate with coal beds； ⑫ Grey arenaceous slate； ⑬ Coarse grained metagreywacker； ⑭ Arenaceous slate； ⑮ Metagreywacker bearing gravels； ⑯ Black slate.

(From figure 2-10 in "The Guide Book for Zhoukoudian Geological Teaching and Practice in Beijing")

太平山煤炭沟奥陶系-石炭系平行不整合剖面

Parallel unconformity between the Ordovician and the Carboniferous in Meitangou of the Taiping Mountain

本溪组硬绿石角岩
Chloritoid hornfels of the Benxi Formation

本溪组（C_2b）：底部普遍发育硬绿泥石角岩及红柱石角岩。下部为杂色粉砂质板岩及变质粉砂岩。中部为含黄铁矿假晶灰色、浅灰色板岩。上部为红柱石角岩。

Chloritoid hornfels and andalusite hornfels are well developed at the bottom of the Benxi Formation (C_2b), which consists of mottled silty slates and metasiltstones in the lower part, grey and grayish slates bearing pyrite pseudomorphs in the middle, and andalusite hornfels in the upper.

压力影板岩
Pressure shadow slates

产于本溪组(C_2b)底部。板岩中为褐铁矿化的黄铁矿假晶。(摄于太平山南坡)

Pressure shadow slates, which occur at the bottom of the Benxi Formation (C_2b), are commonly indicated by pyrites with brownish grey pseudomorphs. (Photographed in the southern slope of the Taiping Mountain)

本溪组底部"三好砾岩"
The "Sanhao" Conglomerates at the bottom of the Benxi Formation

分布于太平山北坡大砾岩山和小砾岩山一带本溪组和马家沟组之间的一套分选好、磨圆好、成分单一(硅质)的砾岩(又称为"三好砾岩")。其与本溪组之间年代从属关系尚未得到确定。(摄于太平山北坡)

The "Sanhao" Conglomerates, which are distributed between the Benxi Formation and the Majiagou Formation in Daliyan Mountain and Xiaoliyan Mountain, are characterized by well sorting, well rounding and simple composition. The relationship in stratigraphy to the Benxi Formation remains unclear. (Photographed in the northern slope of the Taiping Mountain)

本溪组红柱石角岩
Andalusite hornfels in the Benxi Formation

太原组砂岩-板岩露头
Sandstone and shale in the Taiyuan Formation

太原组（C_2-P_1t）：由1～2个沉积旋回组成。旋回的下部主要变质细粒石英砂岩夹灰黑色板岩；上部主要为薄层粉砂岩、板岩、粉砂质板岩。产较丰富植物化石。

The Taiyuan Formation (C_2-P_1t), consisting of one or two sedimentary cycles, is in the lower part of the cycle composed of fine-grained metamorphosed quartz sandstones interlayered with grayish black slates, and in the upper of thin-bedded siltstones, slates and silty slates. Abundant fossil plants also occur in this formation.

山西组（$P_{1-2}s$）中-粗粒岩屑砂岩
Medium-coarse grained lithic sandstones in the Shanxi Formation

山西组（$P_{1-2}s$）：由两个沉积旋回组成。下部旋回底部为褐灰色变质中粗粒岩屑砂岩。向上沉积粒度变小，发育交错层理。旋回上部为黑色炭质板岩夹煤层。上部旋回下部为变质中细粒变质岩屑砂岩，上部为黑色炭质板岩。本组植物化石丰富。（摄于太平山南坡）

The Shanxi Formation ($P_{1-2}s$) consists of two sedimentary cycles and yields rich fossil plants. In the lower cycle, the formation is at the bottom composed of brownish grey medium-coarse grained metamorphosed lithic sandstones with grain getting finer upwards and cross beddings, and in the upper of black carbonaceous slates interlayered with coal beds. While in the upper cycle, exposed in the lower part are fine-medium grained metamorphosed lithic sandstones, and in the upper are black carbonaceous slates. (Photographed in the southern slope of the Taiping Mountain)

杨家屯组底部变质复成分角砾岩
Metamorphosed polymictic conglomerates at the bottom of the Yangjiatun Formation

杨家屯组（P_2y）：由2～3个沉积旋回组成，以粗碎屑沉积为主。旋回下部为灰色变质中-粗粒岩屑砂岩；上部为变质细粒岩屑砂岩、粉砂岩及板岩。（摄于太平山南坡）

The Yangjiatun Formation (P_2y), which consists of two-three sedimentary cycles, is mainly in lithology composed of coarse clastic sediments. Each cycle in the formation is represented in the lower part by grey medium-coarse grained metamorphosed lithic sandstones, and in the upper by fine-grained metamorphosed lithic sandstones, siltstones and slates. (Photographed in the southern slope of the Taiping Mountain)

红庙岭组（$P_{2-3}h$）含砾粗砂岩
Conglomeratic grit in the Hongmiaoling Formation

红庙岭组（$P_{2-3}h$）：由多个沉积旋回组成。旋回下部主要是变质含砾长石石英粗砂岩，向上过渡为变质细砂岩，发育板状交错层理及水流波痕；上部为红色板岩、粉砂质板岩、粉砂岩。（摄于三盆山）

The Hongmiaoling Formation ($P_{2-3}h$), consisting of several sedimentary cycles, is in lithology composed of metamorphosed conglomeratic feldsparthic quartzose grit in the lower part of each cycle that turns to packsand upwards where planer cross beddings and current ripple marks occur; While the upper is composed of red slates, silty slates and siltstones. (Photographed in the Sanpen Mountain)

双泉组（Ts）中-细粒砂岩
Fine-medium grained sandstone in the Shuangquan Formation

双泉组（Ts）：下部变质中细粒砂岩及板岩。上部以灰色中厚层变质细砂岩为主，夹变质粉砂岩及板岩，泥砾发育。本组跨晚二叠世及三叠纪。（摄于三盆山）

The Shuangquan Formation (Ts), which spans Late Permian to Triassic, is in the lower part composed of fine-medium grained metasandstones and slates, and in the upper part of grey medium-thick bedded metapacksands interlayered with metasiltstones and slates where muddy gravels occur. (Photographed in the Sanpen Mountain)

窑坡组(J₁y)含砾碳质、粉砂质板岩

Conglomeratic carbonaceous slate and silty slate in the Yaopo Formation

窑坡组(J₁y)：变质砂岩夹黑色含炭质粉砂质板岩和千枚岩及数层可采煤。（摄于三盆山）

The Yaopo Formation (J₁y) is composed of metasandstones interlayered with black carbonaceous silty slates, phyllite and exploitable coal. (Photographed in the Sanpen Mountain)

龙门组(J₂l)石英岩质变质砾岩

Metamorphosed quartzose conglomerate in the Longmen Formation

龙门组(J₂l)：主要岩性为含炭质粉砂质板岩、千枚状板岩及变质砂岩。（摄于三盆山）

The Longmen Formation (J₂l) consists of carbonaceous silty slates, phyllitic slates and metasandstones. (Photographed in the Sanpen Mountain)

九龙山组(J₂j)变质砾岩

Metamorphosed conglomerate in the Jiulongshan Formation

九龙山组(J₂j)：底部为中-粗粒变质砾岩。下部为变质凝灰质砂岩；中部为变质凝灰质细砂岩夹多层变质砾岩、砂岩、板岩；上部为浅灰色凝灰质砂岩、粉砂岩夹含砾火山岩屑砂岩。（摄于三盆山）

The Jiulongshan Formation (J₂j) at the bottom consists of medium-coarse grained metamorphosed conglomerate, in the lower part of metamorphosed tuff sandstone, in the middle of metamorphosed tuff packsand interlayered with several beds of metasandstones, sandstones and slates, and in the upper of grayish tuff sandstones and siltstones interlayered with conglomeratic volcanic lithic sandstones. (Photographed in the Sanpen Mountain)

2.2 古生物
Paleontology

2.2.1 叠层石
Stromatolites

时代：中元古界
产地：八角寨铁岭组
Age: Middle Proterozoic
Locality: Tieling Formation in Bajiaozhai

2.2.2 古动物化石
Fossil animals

莱德利基虫
Redlichia sp.

节肢动物门，三叶虫纲
时代：晚寒武世
产地：区外
Phylum Arthropoda, Class Trilobita
Age: Late Cambrian
Locality: beyond the practice area

蝴蝶虫
Blackwelderia sp.

节肢动物门，三叶虫纲；时代：晚寒武世；产地：牛口峪。
Phylum Arthropoda, Class Trilobita; Age: Late Cambrian; Locality: Niukouyu.

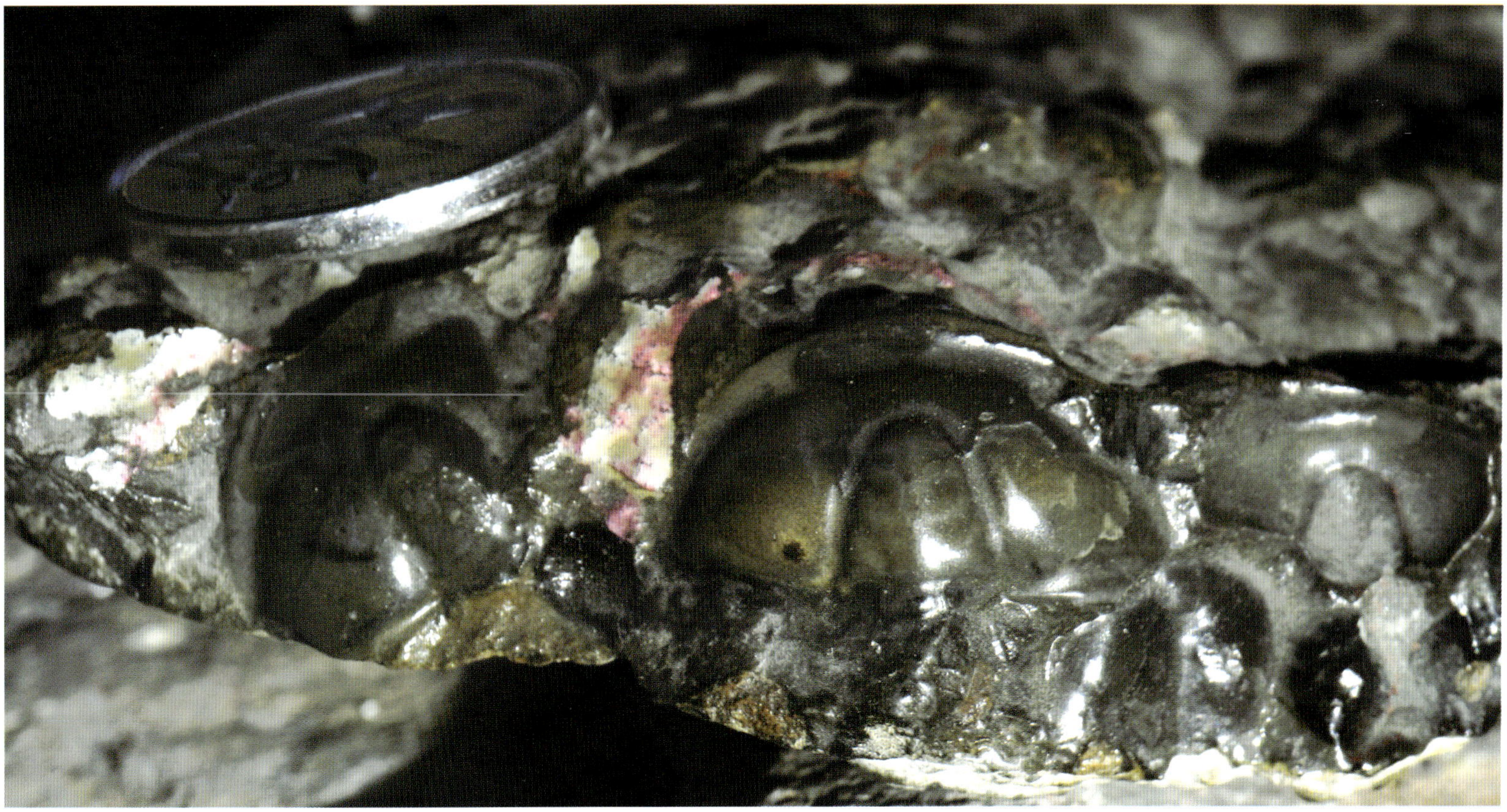

毕雷氏虫
Bailiella matthew

节肢动物门，三叶虫纲
时代：中寒武世
产地：区外
Phylum Arthropoda, Class Trilobita
Age: Middle Cambrian
Locality: beyond the practice area

圆货贝
Obolus sp.

腕足动物门，无铰纲
时代：晚寒武世
产地：区外
Phylum Brachiopoda, Class Inarticulata
Age: Late Cambrian
Locality: beyond the practice area

毕雷氏虫
Bailiella matthew

节肢动物门，三叶虫纲
时代：中寒武世
产地：区外
Phylum Arthropoda, Class Trilobita
Age: Middle Cambrian
Locality: beyond the practice area

刺毛珊瑚
Chaeteties sp.

腕足动物门，珊瑚纲
时代：晚石炭世
产地：太平山煤炭沟本溪组
Phylum Coelenterata, Class Anthozoa
Age: Late Carboniferous
Locality: Benxi Formation in Meitangou of the Taiping Mountain

乳孔贝
Acrothele sp.

腕足动物门，无铰纲
时代：晚寒武世
产地：牛口峪
Phylum Brachiopoda, Class Inarticulata
Age: Late Cambrian
Locality: Niukouyu

2.2.3 古植物化石
Fossil Plants

碳质页岩含植物化石
Carbon shale containing fossil plants

带羊齿
Taeniopyeris sp.

种子蕨植物门
时代：早-中二叠世
产地：太平山山西组
Phylum Pteridospermophyta
Age: Early-Middle Permian
Locality: Shanxi Formation in the Taiping Mountain

芦木
Calamites sp.

节蕨植物门
时代：早-中二叠世
产地：太平山山西组
Phylum Arthrophyta
Age: Early-Middle Permian
Locality: Shanxi Formation in Meitangou of the Taiping Mountain

科达
Cordaites sp.

松柏植物门
时代：早-中二叠世
产地：太平山山西组
Phylum Coniferophyta
Age: Early-Middle Permian
Locality: Shanxi Formation in the Taiping Mountain

瓣轮叶
Lobatannularia sp.

节蕨植物门
时代：早—中二叠世
产地：太平山山西组、杨家屯组
Phylum Arthrophyta
Age: Early-Middle Permian
Locality: Shanxi Formation and Yangjiatun Formation in the Taiping Mountain

脉羊齿
Neuropteris sp.

种子蕨植物门
时代：早—中二叠世
产地：太平山太原组、山西组
Phylum Pteridospermophyta
Age: Late Carboniferous-Early Permian
Locality: Taiyuan Formation and Shanxi Formation in the Taiping Mountain.

3

矿物、岩石
Minerals and Rocks

周口店地区出露的岩石,除燕山期侵入体外,多数遭到了不同程度的变质作用。它们有太古宙变质杂岩、元古宙和古生代区域变质岩、岩体周围的接触热变质岩及与构造变形作用有关的动力变质岩。主要造岩矿物常见斜长石、正长石、角闪石、石英、黑云母、方解石;副矿物常见磁铁矿、磷灰石、黄铁矿、红柱石等。

Rocks outcropped in the Zhoukoudian area mostly experienced metamorphism more or less to some extent, except the Yanshanian intrusions. Major outcropped rocks consist of the metamorphic Complex of Archean, regional metamorphic rocks of Proterozoic-Paleozoic, contact thermal metamorphic rocks around plutons, and dynamic metamorphic rocks associated with tectonics. Common rock-forming minerals include plagioclase, orthoclase, amphibole, quartz, biotite, calcite and accessory minerals include magnetite, apatite, pyrite and andalusite and so on.

3.1 矿物
Minerals

矿物是各种地质作用下形成的天然单质或化合物,具有固定的化学成分和内部结构,一般有一定的形态、物理性质和化学性质。

周口店常见的矿物有两大类:造矿矿物和造岩矿物,本图册介绍的造岩矿物有钾长石、斜长石、黑云母、角闪石、红柱石、石英、方解石,造矿矿物主要有黄铁矿和磁铁矿。

Minerals are naturally constructed by single elements or compounds developed in geological processes, which are composed of fixed chemical compositions and internal structures, and bear certain morphological, physical and chemical features.

Two mineral classes are common in the Zhoukoudian area: ore forming minerals and rock forming minerals. Rock forming minerals introduced in this atlas are K-feldspar, plagioclase, biotite, amphibole, andalusite, quartz and calcite; while ore forming minerals are mainly pyrite and magnetite.

正长石的晶形
The crystal shape of orthoclase

正长石(K[AlSi$_3$O$_8$]单斜晶系)
Orthoclase (K[AlSi$_3$O$_8$], monoclinic crystal)

正长石是钾长石的一个种,晶体常呈短柱状或厚板状。以其肉红色、卡氏双晶、高硬度、两组正交解理作为重要的鉴别标志。

周口店房山复式岩体和伟晶岩脉中存在大量正长石,颗粒粗大,卡氏双晶较多。(摄于周口店西风坡)

Orthoclase is a variety of K-feldspars, which occurs as stubby or blocky-tabular crystals. Identification of orthoclase mainly includes its flesh-red color, Karlsbad twinning, high hardness, and a couple of orthographical cleavages.

In the Zhoukoudian area, orthoclases occurring as coarse grained with Karlsbad twinning, are rich in the Fangshan multi-intruded pluton and related pegmatite veins. (Photographed at Xifengpo in Zhoukoudian)

文象结构
Graphic texture

伟晶岩中的石英呈一定的外形（如尖棱形，象形文字形等）有规律地镶嵌在正长石里形成文象结构。周口店房山复式岩体伟晶岩脉中可见文象结构。（摄于周口店官地村）

In pegmatite veins, quartz is commonly interpenetrated into orthoclases with a certain shape (as arris-shaped, hieroglyph), which is referred as the graphic texture. Graphic textures widely appear in pegmatite veins of the Fangshan multi-intruded pluton. (Photographed at Guandi Village in Zhoukoudian)

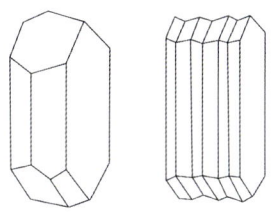

斜长石的晶形（a）
与聚片双晶（b）
The crystal shape of plagioclase(a) and polysynthetic twinning(b)

a　　b

斜长石（Na[AlSi$_3$O$_8$]－Ca[Al$_2$Si$_2$O$_8$]三斜晶系）
Plagioclase (Na[AlSi$_3$O$_8$]-Ca[Al$_2$Si$_2$O$_8$], triclinic crystal)

斜长石具板状及板柱状晶形，集合体常呈板状或不规则粒状，肉眼易见聚片双晶，以其白色、两组完全解理，且交角约94°、86°和高硬度区别于其他相似矿物。

周口店房山复式岩体及变质岩、沉积岩中均有大量斜长石产出，其中岩体中斜长石结晶较好。（摄于周口店西风坡）

Plagioclases occur as tabular or blocky-tabular crystals while the aggregates are commonly tabular or irregular-grainy, in which polysynthetic twining can be nake-eyed. Plagioclases are distinguished with other similar crystals by their white color, a couple of good cleavages with angles of 94 and 86 degrees, and high hardness.

In the Zhoukoudian area, abundant plagioclases appear in the Fangshan multi-intruded pluton, metamorphic and sedimentary rocks, of which igneous plagioclases are euhedral. (Photographed at Xifengpo in Zhoukoudian)

黑云母(K{(Mg,Fe)₃[AlSi₃O₁₀](OH)₂} 单斜晶系)
Biotite (K{(Mg,Fe)₃[AlSi₃O₁₀](OH)₂}, monoclinic crystal)

黑云母晶体呈现假六方板状或锥形短柱状，集合体呈片状或鳞片状，以其黑色、一组极完全解理及低硬度区别于其他相似矿物，是中酸性岩浆岩及变质岩的主要造岩矿物。

周口店黑云母主要见于古老的变质岩、接触变质岩和房山复式岩体、伟晶岩脉中。（摄于周口店官地村西）

Biotitic crystals appear as pseudohexgonal tabular or stubby pyramids and aggregates are flaky. Biotite, which is a common rock-forming mineral in intermediate-acid igneous and metamorphic rocks, is differentiated by its dark color, a set of perfect cleavage and low hardness from other similar minerals.

In the Zhoukoudian area, biotite occurs mainly in old regional metamorphic, contact metamorphic rocks and in the Fangshan multi-intruded pluton as well as pegmatite veins. (Photographed at the west of Guandi Village in Zhoukoudian)

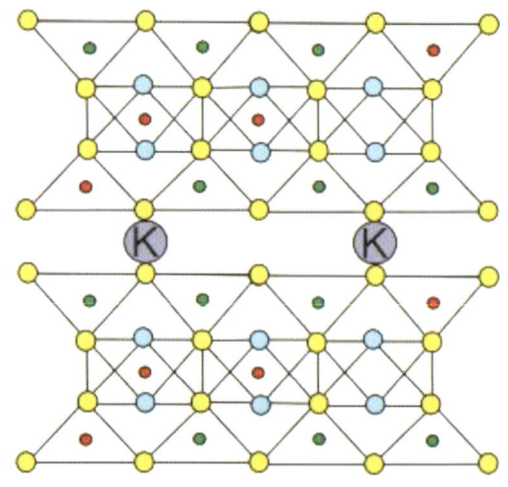

云母的晶体结构图
The crystal structure of Biotite

普通角闪石（$Ca_2Na(Mg,Fe^{2+})_4(Al,Fe^{3+})[(Si,Al)_4O_{11}]_2(OH)_2$ 单斜晶系）
Hornblende ($Ca_2Na(Mg,Fe^{2+})_4(Al,Fe^{3+})[(Si,Al)_4O_{11}]_2(OH)_2$, monoclinic crystal)

普通角闪石晶体呈长柱状，横断面呈菱形或近菱形的六边形，集合体呈细柱状、纤维状，以其柱状形态，六边形横切面，两组夹角56°或124°解理为特征，常见于变质岩和岩浆岩中。

周口店的普通角闪石见于古老的变质岩和房山复式岩体中。
（摄于西风坡东）

Hornblendes are column crystals with hexagonal cross section, of which aggregates are long column or fibroid. Hornblendes that are common in metamorphic and igneous rocks, are characterized by their column crystal shape, hexagonal cross section and a couple of cleavages with angular separations of 54 and 126 degrees.

Hornblendes in the Zhoukoudian area mainly exist in old metamorphic rocks and in the Fangshan multi-intruded pluton. (Photographed at the east of Xifengpo in Zhoukoudian)

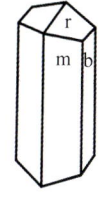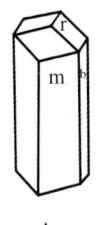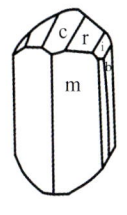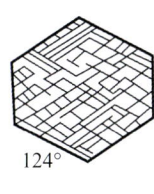

A B

角闪石的晶形（A）与两组解理（B）
The crystal shape of Hornblede(A)
The cleavages in two directions of Hornblede(B)

透闪石（$Ca_2mg_5[Si_4O_{11}]_2(OH)_2$ 单斜晶系）
Tremolite ($Ca_2mg_5[Si_4O_{11}]_2(OH)_2$, monoclinic crystal)

透闪石晶体常呈柱状、纤维状，集合体呈柱状、放射状。与普通角闪石特征近似，只是颜色不同。

周口店的透闪石主要见于房山岩体接触变质带大理岩中。（摄于周口店羊屎沟）

Tremolites commonly occur as columns and fibroid, while aggregates as columns and radial shape. Tremolite is very similar to hornblende except its color.

Tremolites in the Zhoukoudian area mainly exist in marbles developed in the contact metamorphic zone of the Fangshan pluton. (Photographed at Yangshigou in Zhoukoudian)

红柱石（$Al_2[SiO_4]O$ 斜方晶系）
Andalusite ($Al_2[SiO_4]O$, rhombic crystal)

红柱石晶体呈柱状，横断面近正方形。集合体呈柱状或放射状。放射状集合体形似菊花，故又名菊花石。含碳质包裹体的红柱石称为空晶石。空晶石有独特的碳质包裹体，可与其相似的矿物区别。

红柱石分布于周口店房山复式岩体周围的泥质岩地层中，为接触热变质作用形成。（摄于周口店升平山）

Andalusites are column crystals that with square cross section. Aggregates are columns or radial shaped that are similar to chrysanthemums, which is the reason that andalusite is also referred to chrysanthemum stone. Andalusite is also named as chiastolite if including carbonic inclusions, which is differentiated from other similar minerals.

Andaulsites in the Zhoukoudian area mainly occur in thermal contact metamorphic zone in pelitic strata around the Fangshan multi-intruded pluton. (Photographed at Shengping Mountain in Zhoukoudian)

a　　　　　　　　　　b

红柱石的晶形（a）
与其横切面特征（b）
The crystal shape of Andalusite (a)
The cross section feature of Andalusite (b)

石英（SiO_2 α-石英为三方晶系，β-石英为六方晶系）
Quartz (SiO_2 α-trigonal crystal system, β-hexagonal crystal system)

石英以其晶形、油脂光泽、无解理、贝壳状断口、硬度大等特征区别于其他相似矿物。可发育成完好的六方柱状晶体，柱面有横向聚形纹。石英是自然界分布广泛的造岩矿物之一，在内生和外生条件下均可生成。

周口店的石英常见于岩体、岩脉中，官地村伟晶岩脉中可以见到水晶晶体。（摄于周口店空地林西）

Quartz is differentiated according to its crystal shapes, glassy to vitreous luster, no cleavage, conchoidal fractures and high hardness. Perfect hexagonal-column crystals with lateral combination striations are common. Quartz is one of the most common rock forming natural minerals that develops both exogenously and endogenously.

In the Zhoukoudian area, quartz commonly occurs in plutons and veins, e.g. the quartz crystals in pegmatite veins in the Guandi Village. (Photographed at Xifengpo to the west of the Guandi Village in Zhoukoudian)

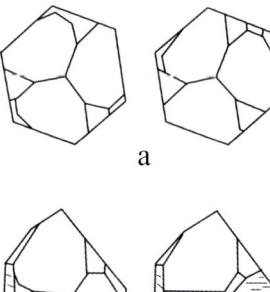

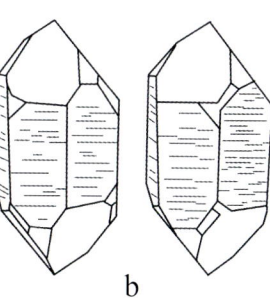

石英的晶形（a）与聚形纹（b）

The crystal shape of Quartz(a)

The combination striations of Quartz(b)

方解石（Ca [CO$_3$]三方晶系）
Calcite (Ca [CO$_3$], trigonal crystal)

方解石常见完好晶体，随温度的不同而表现出不同的形态。集合体常见有晶簇状，致密块状、粒状、土状、多孔状、泉华状、鲕状、钟乳状等。以菱面体完全解理，硬度3，与冷稀 HCl 相遇剧烈起泡等特征区别于其他相似矿物。

周口店各类碳酸盐地层中均可见到。（摄于周口店太平山南坡采石场）

Euhedral calcite crystals are common, which vary with temperatures. Aggregates appear in druse, massive, grainy, claylike, porous, sinter-shaped, oolite or stalactitic forms. Calcites are distinguished with other similar minerals by their perfectly cleavages into rhombohedrons, a hardness of 3 and intense bubbling reaction with diluted hydrochloric acid.

Calcites are common in various carbonatites in the Zhoukoudian area. (Photographed at a quarry in the southern slope of the Taiping Mountain in Zhoukoudian)

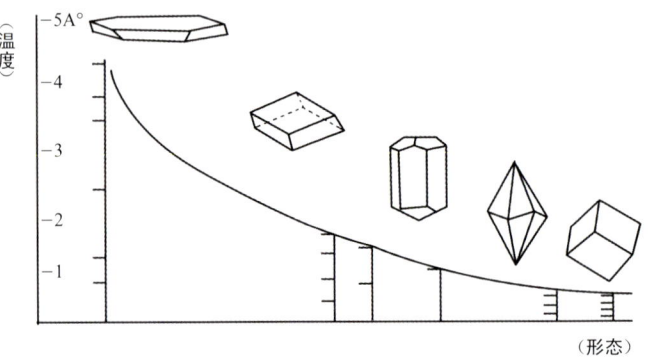

方解石晶形温度降低序列（温度与形态的关系）
Calcite crystal forms that vary with a decreasing temperature sequence

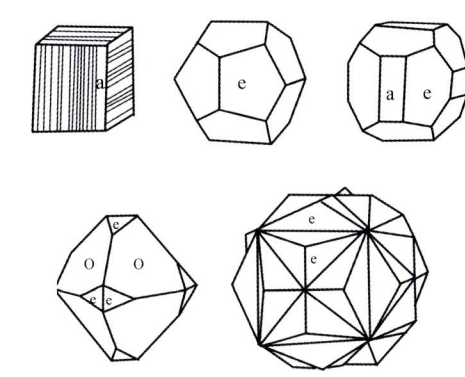

黄铁矿的晶形
The crystal shape of Pyrite

黄铁矿(多数褐铁矿化)(Fe[S_2]等轴晶系)
Pyrite (mostly limonitized) (Fe[S_2], isometric system)

黄铁矿常具有完好的立方体晶形,以其完好的晶形和晶面条纹、颜色、较大的硬度,可与相似的黄铜矿区别。它们的形成必须是还原环境。水解后,形成褐铁矿。

周口店的黄铁矿主要见于石炭—二叠纪地层中,且多数已被褐铁矿化,有的由于应力作用而形成黄铁矿压力影。(摄于周口店太平山南坡)

Pyrites commonly occur as perfect cubic crystals, which are differentiated from chalcopyrites mainly by crystals forms, combination striations, color and higher hardness. Pyrites are formed in reductive conditions and can be hydrolyzed into limonites.

In the Zhoukoudian area, pyrites mainly occur in the Carboniferous-Permian, most of which have been limonitized, and surrounded by pressure shadow structures due to stressing. (Photographed in the southern slope of the Taiping Mountain)

黄铁矿压力影(摄于周口店太平山南坡)
Pressure shadow structure surrounding pyrite (Photographed in the southern slope of the Taiping Mountain in Zhou Koudian)

磁铁矿（FeFe$_2$O$_4$ 等轴晶系）
Magnetite (FeFe$_2$O$_4$, isometric system)

磁铁矿具有八面体或菱形十二面体晶形。集合体通常为致密粒状块体，或分散粒状。以强磁性、黑色条痕可与相似矿物区别。形成于还原环境，岩浆作用，热液作用及变质作用中均可形成。

周口店的磁铁矿具完好的晶形，见于新元古界下马岭组底部含磁铁矿千枚状板岩中。复式岩体中常呈副矿物。（摄于周口店拴马桩）

Magnetite crystals occur in octahedrons or dodecahedrons, while aggregates are massive or single grainy. Magnetites, which form in reductive conditions in igneous and hydrothermal processes as well as metamorphism, are differentiated by magnetism and dark striations from similar minerals.

In the Zhoukoudian area, euhedral magnetites as an accessary mineral in the Fangshan multi-intruded pluton also occur in the magnetite phyllitic slates at the bottom of the Neoproterozoic Xiamaling Formation. (Photographed at Shuanmazhuang in Zhoukoudian)

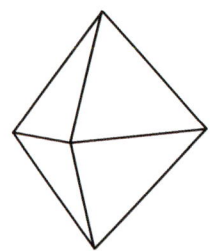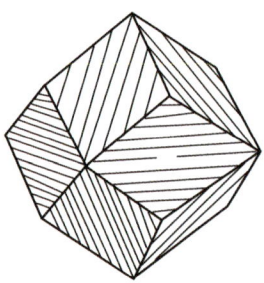

磁铁矿的晶形
The crystal shape of Magnetite

3.2 岩石
Rocks

岩石的种类按其成因可分为三大类：岩浆岩（或称火成岩）、沉积岩、变质岩。三大类岩石并不能截然分开，其间常为逐渐过渡关系，并可互相转化。

周口店地区岩石种类较多，其中，岩浆岩有花岗闪长岩、石英闪长岩、闪长玢岩、伟晶岩脉、煌斑岩等；沉积岩有砾岩、砂岩、白云岩、石灰岩、鲕粒灰岩、竹叶状灰岩、泥质条带灰岩、豹皮灰岩、纹层灰岩、火山碎屑岩、凝灰岩等；变质岩有斜长角闪岩、片麻岩、混合岩、板岩、大理岩、红柱石角岩、各种动力变质岩等。

Rocks can be classified into three types: magmatic (igneous), sedimentary and metamorphic rocks. These types are not completely distinct but are inter-transformable, between which transitional rock types exist.

Various rock types occur in the Zhoukoudian area. Magmatic rocks appear as granodiorite, quartz diorite, diorite－porphyrite, pegmatite veins, lamprophyre and so on; sedimentary rocks occur as conglomerates, sandstones, dolomites, limestones, oolitic limestones, wormkalk limestones, banded pelitic limestones, leopard limestones, lamella limestones, pyroclastic rocks and tuff; while metamorphic rocks as plagioclase amphibolite, gneisses, migmatites, slates, marble, andalusite hornfels and various dynamic metamorphic rocks.

3.2.1 岩浆岩
Magmatic rocks

（石英闪长岩显微镜下特征(+)）
(The feature of Quartz diorite in microscope(+))

花岗闪长岩
Granodiorite

花岗闪长岩中的斜长石多于钾长石，占长石矿物总量的65%～90%，斜长石一般为酸性和中性斜长石。石英含量一般在25%左右。深色矿物角闪石和黑云母，常见半自形粒状结构，似斑状结构。

周口店房山复式岩体被分成三个岩相带，主要由不同结构的花岗闪长岩组成。

Granodiorite mainly consists of feldspar, quartz and dark minerals. The feldspar mainly consists of 65%~90% plagioclase, which is acid to intermediate, and of miner K-feldspar. Quartz in granodiorite amounts to about 25%. The dark minerals are composed of hornblende and biotite with hypidiomorphic-granular texture and porphyritic-like texture.

Three lithofacies have been identified in the Fangshan multi-intruded pluton, which is constructed by granodiorites of different textures.

石英闪长岩
Quartz diorite

石英闪长岩主要由斜长石、石英和钾长石镁铁质矿物组成。石英含量为 5%～20%，钾长石含量在 10%～15%，暗色矿物含量在 15% 左右，斜长石（中长石）占一半以上，岩石具半自形粒状结构。

周口店房山复式岩体石英闪长岩作为房山花岗闪长岩体边缘相和早期第一次侵入产物分布在岩体的边部。

Quartz diorite, with hypidiomorphic-granular texture, mainly consists of plagioclase, quartz, K-feldspar and mafic minerals. Quartz amounts to 5%~20%, K-feldspar amounts to 10%~15%, dark minerals amounts to 15% and plagioclase to 50%.

Quartz diorite occurs as marginal facies in the earliest intrusion phase, which is located at the edge of the Fangshan multi-intruded pluton.

（显微镜下的石英闪长岩（+））
（Quartz diorite in microscope（+））

闪长玢岩
Diorite-porphyrite

闪长玢岩属中性浅成侵入岩石，石英大于 5%，暗色矿物 20%～35%，长石类矿物主要为中性斜长石（中长石），斑状结构；不含或仅含少量碱性长石；最常见的暗色矿物为角闪石，也有以黑云母或辉石为主者。

此类岩脉主要分布于周口店太平山北坡。（摄于周口店大砺岩山北采石坑）

Diorite-porphyrite is an intermediate shallow-intrusive rock with quartz>5% and dark minerals 20%~30%. Feldspar minerals mainly consist of intermediate plagioclase with porphyritic structure and rarely of alkaline plagioclase. The most common dark minerals are hornblende, and also biotite and pyroxene.

Related veins are mainly located in the southern slope of the Taiping Mountain in Zhoukoudian. (Photographed at a quarry to the north of the Daliyan Mountain in Zhoukoudian)

伟晶岩
Pegmatite

伟晶岩主要矿物组成简单，有石英、碱性长石和斜长石，白云母、黑云母等。矿物颗粒粗大，具伟晶结构、文象结构、晶洞、晶线构造。

伟晶岩作为岩脉分布于房山复式岩体中。（摄于官地村西 127.2 高地）

Pegmatite with simple mineral composition consists of quartz, alkali feldspar, plagioclase, muscovite and biotite and so on, which are coarse grained with pegmatitic texture, graphic texture and miarolitic structure.

Pegmatites occur as veins and widely distribute in the Fangshan multi-intruded pluton. (Photographed at the 127.2 high point to the west of the Guandi Village)

煌斑岩脉（穿插在164背斜南翼奥陶纪地层中）

Lamprophyres (intruded in the Ordovician at the southern limb of the highpoint 164 anticline)

煌斑岩脉，化学成分多数接近基性岩，少数接近超基性岩。矿物成分黑云母和角闪石含量最高，其次是辉石、橄榄石很少。浅色矿物以钾长石和斜长石为主，也可有似长石。岩石多为斑状结构、粒状结构，暗色矿物多呈自形晶和半自形斑晶称为煌斑结构。具斑状结构的煌斑岩，斑晶几乎全为自形程度高的暗色矿物。煌斑岩类容易风化和蚀变，常见有绿泥石化，碳酸岩化和高岭土化等。（摄于太平山南坡采石场）

Lamprophyres, which are mostly close to mafic rocks in chemistry, and minor to ultra mafic rocks, consist of dark minerals of biotite and hornblende, and very minor pyroxene and olivine, and light-colored minerals mainly of K-feldspar, plagioclase and also feldspathoid. Rocks commonly occur as porphyritic structure, granular texture and lamprophyric texture, in which dark minerals are euhedral to subhedral porphyritic crystals. In porphyritic structured lamprophyres, porphyritic crystals essentially consist of dark euhedral minerals. Field lamprophyres are commonly weathered and altered, which show chloritization, carbonation, kolinization and so on. (Photographed at a quarry in the southern slope of the Taiping Mountain in Zhoukoudian)

煌斑岩脉（破窑山采石场）

Lamprophyres (a quarry of Poyao Mountain)

包体
Enclave

在岩浆岩中，可以存在各种岩石包体，与其寄生的岩浆岩相比，包体颜色深、粒度细、主要为闪长质成分。（摄于官地村西 125.5 高地）

Various enclaves occurring in the magmatic rock, which are mainly dioritic in composition, have darker color and finer grain size compared to wall rocks. (Photographed at the 125.5 high point to the west of the Guandi Village)

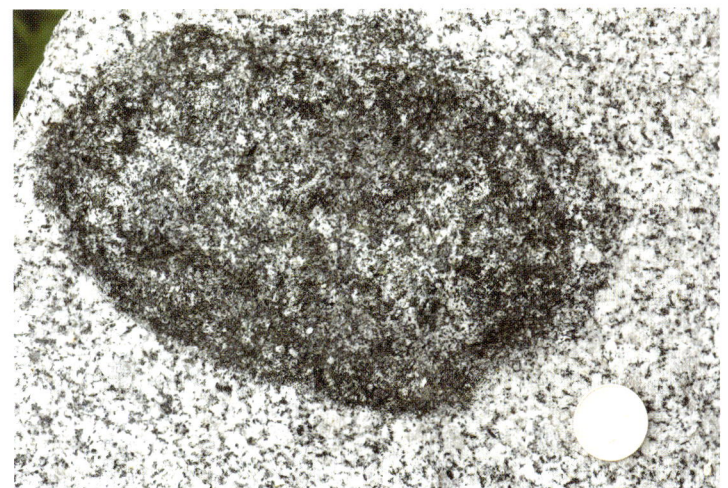 

3.2.2 沉积岩
Sedimentary rocks

石灰岩
Limestone

以方解石为主要成分的碳酸盐岩简称灰岩。有时含有白云石、粘土矿物和碎屑矿物,有灰、灰白、灰黑、黄、浅红、褐红等色,硬度一般不大,与稀盐酸反应剧烈。当粘土矿物含量达 25%～50%时,称为泥质灰岩。白云石含量达 25%～50%时,称为白云质灰岩。

周口店石灰岩主要分布于寒武系和奥陶系地层中。(摄于周口店太平山奥陶系马家沟组地层中)

Limestones refer to a carbonatite rock, which is essentially composed of calcite, and occasionally some dolomite, clay minerals and detritus. Limestones are usually gray, greyish white, yellow, light red, red brown in color, and moderate in hardness, and are intense in reaction with diluted hydrochloric acid. Those when clay minerals and dolomites amount to 25%~50% are named as argillaceous limestone and dolomitic limestone respectively.

Limestones in the Zhoukoudian area mainly occur in the Cambrian and Ordovician. (Photographed from the Majiagou Formation of the Ordovician in the Taiping Mountain in Zhou Koudian)

纹层灰岩
Lamella limestone

纹层灰岩是具有纹层构造的灰岩,纹层通常由藻类骨架构成。见于周口店寒武系府君山中。(摄于黄院东山梁)

Lamella limestones are limestones with lamella structure that is constructed by algae skeleton. The shown occurs in the Fujunshan Formation of the Cambrian, Zhoukoudian. (Photographed at the eastern ridge of Huangyuan)

豹皮灰岩
Leopard limestone

为不规则斑纹的石灰岩，貌似豹皮，故名豹皮灰岩。通常基质部分为隐晶质方解石或微晶方解石，斑纹部分含有较多的白云石。它是石灰岩在成岩过程中发生白云石化而成的，白云石化作用常选择石灰岩中渗透性较好含颗粒的条带或斑块进行。见于周口店寒武系府君山组。（摄于黄院东山梁）

Leopard limestone is named after its irregular strips that are similar to leopard skin. Matrix in leopard limestone is composed of cryptocrystalline or microcrystal calcite, while strips are mainly of dolomitic composition. Leopard limestone forms by dolomitization during diagenetic process of limestone, which preferentially selects those strips in limestone that are of grainy structure with good penetration. Leopard limestones occur in the Fujunshan Formation of Cambrian in Zhoukoudian. (Photographed at the eastern ridge of Huangyuan)

（显微镜下的豹皮灰岩（+））
（Leopard limestone in microscope（+））

鲕粒灰岩
Oolitic limestone

鲕粒灰岩是一种以鲕粒为主要组分的石灰岩。它是兼具化学和机械成因的石灰岩，形成于碳酸钙处于过饱和状态的海、湖波浪活动地带或潮汐通道水流活动地带。多见于周口店寒武系地层中。（摄于周口店黄院东山梁）

Oolitic limestone is a kind of limestone rich in oolites, which is of both chemical and mechanical genesis. It occurs in ocean and lakes with wave activity, and channels with tidy flowing, where calcium carbonates are super-saturated. Oolitic limestones are mostly shown in the Cambrian in Zhoukoudian. (Photographed at the eastern ridge of Huangyuan in Zhou Koudian)

（显微镜下的鲕粒灰岩（+））
(Oolitic limestone in microscope（+）)

竹叶状灰岩
Leafy limestone

石灰岩的一种，其特点为截面有砾石呈竹叶状。它的形成是碳酸盐碎屑经海水常年冲击、侵蚀，慢慢变成类似橄榄状碎石块，一般长 0.3～10 厘米，这种碎屑被钙质胶接、粘合、挤压在一起而固结成岩。见于周口店寒武系-奥陶系地层中。（摄于周口店实习站陈列室）

A kind of limestone features in leafy shaped conglomerates. Leafy limestone forms by calcite-cemented leafy-shaped conglomerates of 0.3~10 cm, which are derived from carbonatite stones crushed and weathered by long lasting waves. Leafy limestones occur in the Cambrian-Ordovician in Zhoukoudian. (Photographed in the geology showroom of the station)

（显微镜下的竹叶状灰岩(+)）
(Leafy limestone in microscope(+))

泥质条带灰岩
Banded pelitic limestone

泥质条带灰岩是薄层状灰岩与泥质岩互层状排列，反映了地壳的不断运动。见于周口店寒武系地层，以黄院组居多。（摄于黄院东山梁）

Banded pelitic limestone consists of interbedded thin limestones and pelitic beds, which may imply an unstable crustal movement. Banded pelitic limestone occurs in the Cambrian, and mostly in the Huangyuan Formation in Zhoukoudian. (Photographed at the eastern ridge of Huangyuan)

（显微镜下的泥质条带灰岩（+））
(Banded pelitic limestone in microscope（+）)

白云岩
Dolomite

白云岩是以白云石为主要组分的碳酸盐岩石，含有少量的方解石和黏土等矿物，主要成分为碳酸镁钙和少量的二氧化硅、氧化铁、氧化铝等。外表类似石灰岩，滴上盐酸不发泡或微弱发泡，风化后有"刀砍纹"。

周口店地区以中元古界地层中大量出现，寒武系和奥陶系也有。（摄于黄院东山梁）

As a kind of carbonatite stone, dolomite mainly consists of dolomite, minor calcite and clay minerals. In chemistry, dolomite is composed of calcium magnesium carbonate, minor silicon dioxide, iron oxide, aluminium oxide and so on. Though dolomite has an appearance like limestone, little or no bubbling reaction occurs with diluted hydrochloric acid. In addition, weathered dolomite has chop profiles.

In the Zhoukoudian area, dolomite appears in mass in the Mesoproterozoic, and also some in the Cambrian and Ordovician. (Photographed at the eastern ridge of Huangyuan)

（显微镜下的白云岩（+））
(Dolomlte in microscope（+）)

石英岩质砾岩(三好砾岩)
Quartzose conglomerates (The "Sanhao" Conglomerate)

石英岩质砾岩的砾石成分简单，主要为稳定性较高的石英岩，脉石英、燧石等；砾岩的磨圆和分选均较好；砾石直径多小于50毫米。胶结物以硅质为主，多呈颗粒胶结；砂、粉砂和粘土组成的填隙较少。由于颗粒的分选、磨圆及成分的单一，为便于理解记忆常称作"三好砾岩"。（摄于周口店太平山北本溪组底部）

The conglomerates are simple in composition, of which gravels are mainly composed of stable quartzite, vein quartz, chert and so on. Gravels are well sorted and rounded with diameters less than 50 mm mostly, which are mainly siliceous-cemented with grainy structure, and thus less sandy and slity interstitial material. For an impression of the above features of well sorting, well rounding and simple composition, they are specially named as "Sanhao" conglomerates. (Photographed from the bottom of the Carboniferous Benxi Formation at the south of the Taiping Mountain in Zhoukoudian)

复成分角砾岩(豆腐块砾岩)
Polymictic conglomerate (Tofu shaped conglomerate)

复成分角砾岩的主要成分是长石碎屑颗粒，含量大于50%，多呈白色棱角状，分选差，属陆源碎屑沉积。（摄于周口店太平山南坡二叠系杨家屯组(P_2y)）

Ploymictic conglomerates are terrigenous clastic sediments, which mainly consist of feldspar clasts that amount to 50%, and occur as light-colored, angular and poor sorted. (Photographed from the Permian Yangjiatun Formation (P_2y) in the southern slope of the Taiping Mountain in Zhoukoudian)

砂岩
Sandstone

砂岩由石英、长石和岩屑颗粒构成，具砂质结构，通常是硅质、钙质、泥质胶结。

周口店砂岩见于新元古界下马岭组和龙山组以及石炭系、二叠系、侏罗系和白垩系地层中，有粗、中、细和粉砂岩。（摄于太平山南坡二叠系地层中）

Consisting of quartz, feldspar and detritus, sandstone with sandy structure is commonly cemented by siliceous, calcium or pelitic material.

Sandstones in the Zhoukoudian area appear in the Xiamaling Formation and Longshan Formation of Neoproterozoic, and in the Carboniferous, Permian, Jurassic and Cretaceous, which occur as coarse, medium-and fine-grained and silty sandstones. (Photographed from the Permian in the southern slope of the Taiping Mountain)

火山角砾岩
Volcanic breccia

火山角砾岩是一种火山碎屑岩。主要由粒径为 2～64 毫米的火山角砾组成，也含有其他岩石的角砾及少量的石英、长石等矿物晶屑。多数具明显的棱角，分选差，大小不等。填隙物是火山灰，火山尘。见于周口店侏罗系地层中。（摄于周口店猫儿山）

As a kind of volcanic clastic sediment, volcanic breccia consists essentially of volcanic clasts sized 2~62 mm, and minor crystal fragments of quartz, feldspar and so on. Clasts are clearly angular-shaped and poor sorted, which are interstitially cemented by volcanic tuff and dust. The volcanic breccia occurs in the Jurassic of Zhoukoudian. (Photographed in the Maoer Mountain in Zhoukoudian)

火山凝灰岩
Volcanic tuff

火山凝灰岩是一种火山碎屑岩。主要由粒径小于 2 毫米的晶屑、岩屑及玻屑组成。碎屑物质小于 50%，分选很差，填隙物是更细的火山微尘。（摄于周口店猫儿山侏罗系地层中）

As a kind of volcanic clastic sediment, volcanic tuff consists of crystal fragments, volcanic clasts and glass fragments less than 2 mm in diameter. Poor sorted clasts amount to less than 50%, while interstitial is finer volcanic dust. (Photographed from the Jurassic in the Maoer Mountain in Zhoukoudian)

3.2.3 变质岩
Metamorphic rocks

（显微镜下的斜长片麻岩（+））
(Plagioclase gneiss in microscope(+))

角闪斜长片麻岩
Hornblende-plagioclase gneiss

角闪斜长片麻岩是一种具片麻状构造，矿物成分主要由石英，斜长石及一定量的片状、柱状矿物组成的岩石。暗色矿物主要是角闪石、黑云母。为周口店地区最古老的变质岩之一。（摄于周口店官地村东）

The hornblende-plagioclase gneiss with gneissic structure, consists mainly of quartz, plagioclase and flaky and column minerals, in which common dark minerals are hornblende and biotite.It is one of the most ancient metamorphic rocks occurred in the Zhoukoudian area. (Photographed to the east of the Guandi Village in Zhoukoudian)

片麻岩
Gneiss

片麻岩具片麻状构造，主要由石英，长石及一定量的片状、柱状矿物组成的岩石。一般长石和石英的含量大于70%，暗色矿物小于30%。暗色矿物主要是云母、角闪石。常为中粗鳞片粒状变晶结构。为周口店地区最古老的变质岩之一。（摄于周口店官地村东）

This rock occurs as gneissic structure and consists mainly of quartz, feldspar and some flaky and column minerals. In general, light-colored minerals as quartz and feldspar amount to over 70%, while dark minerals mainly as mica and hornblende are less than 30%. The gneiss commonly has mediate-coarse lepido granoblastic texture. It is one of the most ancient metamorphic rocks occurred in the Zhoukoudian area. (Photographed to the east of the Guandi Village in Zhoukoudian)

混合岩
Migmatite

混合岩由混合岩化作用形成，由基体和脉体两个基本组成部分组成。基体是角闪岩相或麻粒岩相变质岩，代表混合原岩，或多或少受到改造，又称古成体，脉体是长英质或花岗质物质，代表混合岩中新生部分，又称新成体。常见类型呈条带状混合岩。见于周口店太古界官地杂岩中。（摄于周口店山顶庙东）

Migmatites consist of substrate and vein material formed by migmatization. The substrate represents for the protolithic that have been more or less transformed, which are now metamorphic rocks of amphibolite or granulite facies; while the vein material that is felsic or granitic matter, represents for the newly created in migmatization, which is also named as the neosome. Migmatite commonly occurs with banded structure. Migmatite occurs in the Guandi Complex of the Archean in Zhoukoudian. (Photographed to the east of Shandingmiao in Zhoukoudian)

（显微镜下板岩的斑点状板岩（+））
(Spotted slate of slate in microscope (+))

板岩
Slate

板岩是低级变质岩，板状构造，变余泥质结构，具千枚状板状构造者为千枚状板岩。板岩的颜色随其所含有的杂质不同而变化；含钙的遇盐酸会起泡，因此一般以其颜色命名分类，如灰绿色板岩、黑色板岩、钙质板岩等。见于周口店元古代、古生代及中生代地层中，其中寒武纪景儿峪组钙质板岩是良好的装饰材料。（摄于黄院东山梁）

Slates are low grade metamorphic rocks with slaty structure and blastopelitic texture. Slates with phyllitic-slaty structure are named as phyllitic slates. Slates occur in a variety of colors along with the change of the impurity, thus there are gray green slates, black slates and so on. Calcium slates have bubbling reaction with hydrochloric acid. Slates occur widely in the Proterozoic, Paleozoic and Mesozoic, of which calcium slates in the Cambrian Jingeryu Formation are excellent for decoration. (Photographed at the eastern ridge of Huangyuan in Zhoukoudian)

大理岩
Marble

大理岩是主要由方解石或白云石组成的岩石,碳酸盐矿物含量大于50%,具粒状变晶结构,块状或条带状构造。受房山岩体的影响,见于岩体周围碳酸盐岩地层中。(摄于周口店黄院东山梁新元古界景儿峪组)

Marble mainly consists of calcite and dolomite, in which carbonate minerals amount to over 50%. It occurs as granoblastic texture and massive or banded structure. Dominated by the Fangshan intrusion, marble in the Zhoukoudian area occurs in carbonatite strata surrounding the pluton. (Photographed from the Neoproterozoic Jingeryu Formation at the eastern ridge of Huangyuan in Zhoukoudian)

(显微镜下的大理岩(+))
(Marble in microscope (+))

红柱石角岩
Andalusite hornfels

红柱石角岩由中低温热接触变质作用形成。为斑状变晶结构，基质具细粒状变晶结构，放射状构造。原岩为铝质泥岩，主要矿物有石英、红柱石等。受房山复式岩体的影响，分布在岩体稍远周围的泥质变质岩中。（摄于实习站陈列室）

Andalusite hornfels are a product of medium-low thermal contact metamorphism, which have porphyroblastic texture and radial structure, while matrix is with fine granoblastic texture. Andalusite hornfels, of which the protolith is aluminous pelite, mainly consist of quartz and andalusite. Dominated by multi-intrusion of the Fangshan pluton, andalusite hornfels occur in the pelitic metamorphic rocks relatively distant from the surrounding pluton. (Photographed in the geology showroom of the station in Zhoukoudian)

糜棱岩
Mylonite

糜棱岩具糜棱结构的动力变质岩。糜棱岩具 S-C 组构，是粗糜棱岩的特征。为周口店地区最老的变质岩（官地杂岩）之一。（摄于周口店官地村东）

Mylonites occurring as mylonitic texture are dynamic metamorphic rocks, of which the protomylonite features in S-C fabrics. The mylonite (in the Guandi Complex) is one of the oldest metamorphic rocks in the Zhoukoudian area. (Photographed to the east of the Guandi Village in Zhoukoudian)

（显微镜下的长英质糜棱岩(+)）
(Felsic mylonite in microscope (+))

4 构造地质
Structure Geology

　　周口店及邻区大地构造单元属于华北陆块燕山板内（陆内）造山带，在长期演化形成稳定陆块的基础上，后期又被改造而成为活动区。正因为独特的大地构造位置和漫长的地质演化历史，使其不仅保存有不同阶段较为完整的地质事件记录，而且形成了丰富多彩、类型齐全、典型直观且颇具研究意义的各种地质构造现象，共同组合呈现出一幅复杂的地质构造图像。

　　The Zhoukoudian and its adjacent region belong to the Yanshan Orogen (intraplate) of the North China block in tectonics, which had been transformed into tectonically active region from the stable continental block with long-term evolution. The specific tectonic location and the long-term geological history allow records of detailed geological events and preservations of rich and colorful, manifold, typical and intuitionistic geological structures with research significances, all of which combine into an intricate image with various geological structures.

周口店地区地质构造纲要图
Geological structure outline map of the Zhoukoudian area

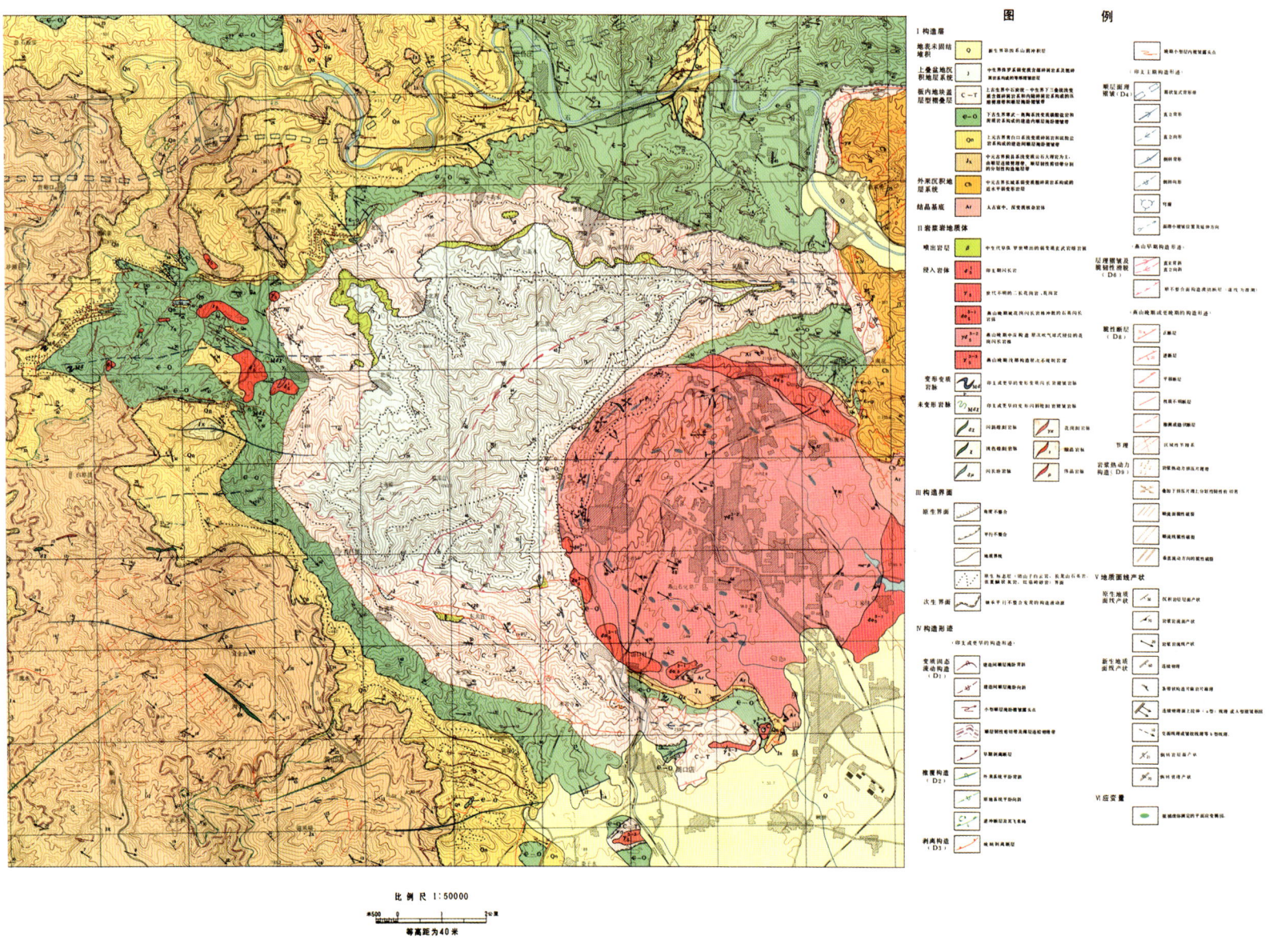

4.1 劈理
Cleavages

　　劈理是一种将岩石按一定方向分割成平行密集的薄片或薄板的次生面状构造，因而使岩石具有了潜在的可劈性。根据其结构和成因，一般可将劈理分为破劈理、滑劈理和流劈理三种类型。

　　Cleavages are a secondary planar structure, which potentially cut rocks into numerous parallel and serried slices with preferential orientation. Based on their internal structures and genesis, cleavages are generally classified into three types: fracture cleavages, slip cleavages and flow cleavages.

破劈理
Fracture cleavages

　　破劈理：是岩石中发育的一组密集的剪切破裂面，破裂面定向与岩石中矿物的排列无关。其与节理的区别，在于密集程度和平行排列程度。如左图即为产于孤山口次级褶皱中的扇状破劈理。结构面具剪-张性，并为后期褐色矿物质充填，呈扇状分布于褶皱转折端。（摄于周口店孤山口）

　　Fracture cleavages are a set of serried shear fractures in rocks, of which the preferential orientation is unrelated to mineral alignment within rocks. They are differentiated from joints mainly in the spatial intensity and the developed extent of the parallel potential planar structure. The left figure shows fan shaped fracture cleavages developed in a secondary fold in Gushankou. (Photographed in Gushankou in Zhoukoudian)

滑劈理
Slip cleavages

　　滑劈理：又称应变滑劈理或褶劈理。常发育在先存面理的岩石中，如板岩、千枚状板岩等，是一组切过先存面理的差异性平行滑动面或滑动带，从而造成先存面理由于差异性平行滑动而褶皱。（摄于周口店孤山口）

　　Slip cleavages, which are also referred to as strain-slip cleavages and crenulation cleavages, are commonly developed in rocks with pre-existing foliations, such as slates and phyllitic slates. They are a set of parallel slip surfaces or zones cutting pre-existing foliations that are dragged and folded by differential slipping along slip cleavages. (Photographed in Gushankou in Zhoukoudian)

流劈理
Flow cleavages

流劈理：流劈理在概念上包括板劈理，是变质岩和强烈变形岩石中常见的一种透入性面理。泛指岩石在固态流变过程中，岩石内部组分发生压扁、拉长、旋转、重结晶并定向排列的现象。左图即为发育在八角寨泥质白云岩褶皱中的流劈理。（摄于周口店八角寨）

Flow cleavages, such as slate cleavages, are a set of penetrative foliations commonly shown in metamorphic and intensively deformed rocks. Generally they are related to material flow within rocks under solid state, including mineral/rock flattening, elongation, rotation, recrystallization and alignment with a preferential orientation. The left figure shows flow cleavages developed in silt dolomites in Bajiaozhai. (Photographed in Bajiaozhai in Zhoukoudian)

轴面劈理
Axial plane cleavages

轴面劈理：是指其产状平行或大致平行于褶皱轴面的劈理（左下，标数字 20 处—正扇状，劈理呈正扇状向褶皱核部收敛；右下薄处—反扇状，劈理呈反扇状向褶皱转折端收敛）。（摄于周口店孤山口）

Axial plane cleavages are referred to as those with attitudes parallel to the axial plane of a fold. The left figure shows a complex fold with secondary folds and axial plane cleavages (Left bottom shows fan-shaped axial cleavages that are convergent to the fold core; while the right bottom shows inverted fan-shaped axial cleavages that are divergent). (Photographed in Gushankou in Zhoukoudian)

层间劈理
Interlayer cleavages

层间劈理：是一种受岩性及层面控制，与层理呈锐角相交的劈理。中元古界铁岭组泥质白云岩岩层中发育的层间劈理，据其与层理之间的锐夹角关系，可判定应力环境为逆时针剪切。（摄于周口店八角寨）

Interlayer cleavages, which generally have acute angles to beddings, are dominated by related lithology and beddings. The figure below shows interlayer cleavages developed in silt dolomites in the Tieling Formation of Mesoproterozoic. Acute angles between the cleavages and beddings imply an anticlockwise shear stress along beddings. (Photographed in Bajiaozhai in Zhoukoudian)

顺层劈理
Bedding cleavages

顺层劈理，是指在宏观上与岩层层理近于平行的劈理。左图即为发育在铁岭组白云岩中的顺层劈理。（摄于周口店八角寨）

Bedding cleavages are parallel or subparallel to beddings in general. The left figure shows bedding cleavages developed in dolomites of the Tieling Formation. (Photographed in Bajiaozhai in Zhoukoudian)

劈理折射
Cleavage refraction

劈理折射由于岩石能干性或岩层厚度的差异，层间发育的劈理与岩层面之间的夹角会出现不同角度的变化，从而造成劈理折射现象。（摄于周口店孤山口）

Cleavage refraction is developed in multilayers with different thickness and/or competence, where cleavages refract when transmitting layers with angle changing between cleavages and beddings. (Photographed in Gushankou in Zhoukoudian)

4.3 线理
Lineations

线理是岩石中广泛发育的一种线状构造,优选方位十分明显。根据尺度可相对划分为大型线理和小型线理;根据其与物质运动方向可划分为 A 型线理和 B 型线理。

Lineations are a linear structure widely developed in rocks with an evident preferential orientation. They can be classified into large-scale and small-scale lineations based on their spatial scale, and into A-typed and B-typed lineations based on their orientation relations to material movement.

窗棂构造
Mullion structure

窗棂构造(B 型线理)发育在能干层与非能干层组合的岩层中,常沿着强弱岩层相邻的强硬层的界面形成一系列形似窗棂的半圆柱状大型线状构造。多由宽而圆的背形被尖而窄的向形所分开,形成嵌入式褶皱组合。(摄于孤山口火车站)

Mullion structure (B-typed lineation) commonly appears in combined layers with competent and incompetent rocks, where competent layer is folded into a series of half-cylinder shell that is similar to a mullion as a large-scale lineation. Mullion structures are linear folds with wide and gentle antiform, and sharp and tight synform. (Photographed in Gushankou in Zhoukoudian)

石香肠构造

Boudinage structure

石香肠构造（B 型线理）是由不同力学性质互层的岩系受到垂直岩层挤压时，发生顺层伸展而形成。在被拉断的强硬层的间隔中，由软弱层呈褶皱状楔入，构成断面上形态各异，顺层展布的不同块体，状似香肠，即称石香肠构造。图即为灰岩中由方解石脉被拉断后形成的藕节状石香肠构造。（周口店实习站地质标本照）

Boudinage structure (B-typed lineation) is commonly formed in competent layers interlayered by incompetent layers, which has been extended and snapped along beddings by normal compressive stress. As a result of injection of soft material into snapped competent layer, the cross section of these layers shows snapped blocks that are similar to boudinage. The figure shows boudinaged calcite veins in limestones. (Photographed in the Base)

第四部分 构造地质

鱼嘴状石香肠构造
Fishhead-shaped boudinage structure

雾迷山组燧石条带白云岩中由燧石构成的鱼嘴状石香肠构造。（摄于周口店十渡景区）
Fishhead-shaped boudinage structures are indicated by chert developed in banded chert dolomites of the Wumishan Formation. (Photographed at the Shidu Scenic Spot)

灯笼状石香肠构造
Lantern-shaped boudinage structure

雾迷山组白云岩中的灯笼状石香肠构造。（摄于周口店孤山口）
Lantern-shaped boudinage structure is developed in dolomites of the Wumishan Formation. (Photographed in Gushankou in Zhoukoudian)

压力影构造
Pressure shadow structure

压力影构造（A 型线理）是矿物生长线理的另一种表现，常产出于低级变质岩中。在应力作用下，相对刚性的矿物黄铁矿、磁铁矿等在变形时可引起局部不均匀应变，造成高压区易溶物质向低压拉张区迁移，沿最大拉伸方向生长成纤维状的影中矿物。（摄于周口店太平山南坡）

Pressure shadow structure (A-typed lineation), which is an equivalent indication of mineral growth lineation, commonly appears in low-grade metamorphic rocks. Within stressed rocks, inhomogeneous strain deformation occurs around competent minerals such as pyrites, magnetites and so on, where soluble substances migrate to low stressed area from high stressed, and fibrous minerals grow along the extensional axis in the strain shadow. (Photographed in the southern slope of the Taiping Mountain)

矿物生长线理
Mineral growth lineations

矿物生长线理（A 型线理）是由针状、柱状矿物在应力作用下沿拉张方向重结晶定向生长的结果。右图为剪节理面上定向生长的纤维状石英。（摄于周口店车厂）

Mineral growth lineations (A-typed lineation) are indicated by needle or prismatic minerals, which recrystallize and grow with preferential orientation along extensional axis under stresses. The right figure shows fibrous quartz growing with preferential orientation on a shear joint. (Photographed in Chechang in Zhoukoudian)

交面线理
Intersection lineations

交面线理（B 型线理）是两组面理相交或面理与层理相交形成的线理。（摄于周口店黄院东山梁）

Intersection lineations (B-typed lineation) are intersections of either two sets of foliations or foliation and bedding. (Photographed in the eastern ridge of Huangyuan in Zhoukoudian)

4.4 褶皱
Folds

  褶皱是由岩层中的各种面（层理、面理等）的弯曲而显示的变形。它规模大小不一、形态复杂多样，形象地反映了地壳岩层所发生的连续变形。是地壳中最普遍的构造现象之一。

  Folds are one of the most common structures in earth crust, of which deformations are indicated by bending of various layers (beddings, foliations and so on). Folds occur in various spatial scales and in abundant configurations, which visualize the continuous deformation of curst rock layers.

拒马河六渡背斜褶皱构造
The Juma River anticline in Liudu

周口店 164 背斜

The highpoint 164 anticline in Zhoukoudian

164 背斜(因位于原 164 高地而得名)平行于其北侧的太平山(天线处山梁)向斜呈东西向展布,并向东倾伏。两翼地层由石炭系—二叠系组成,相背倾斜;核部地层由奥陶纪下奥陶统(O_1m)石灰岩组成,因而被作为石灰、水泥材料大量开采,现仅存其向东倾伏部分。

The highpoint 164 anticline names after its location at the previous 164 highpoint. Plunging to the east, the anticline is located to the south of the Taiping Mountain where the Taiping Mountain syncline occurs parallel. The anticline limbs are constructed by the Carboniferous-Permian, and the core is constructed by limestones of Lower Ordovician (O_1m), which have been largely explored for producing quicklime and cement. So far only preserves the east plunging part of the anticline.

第四部分 构造地质

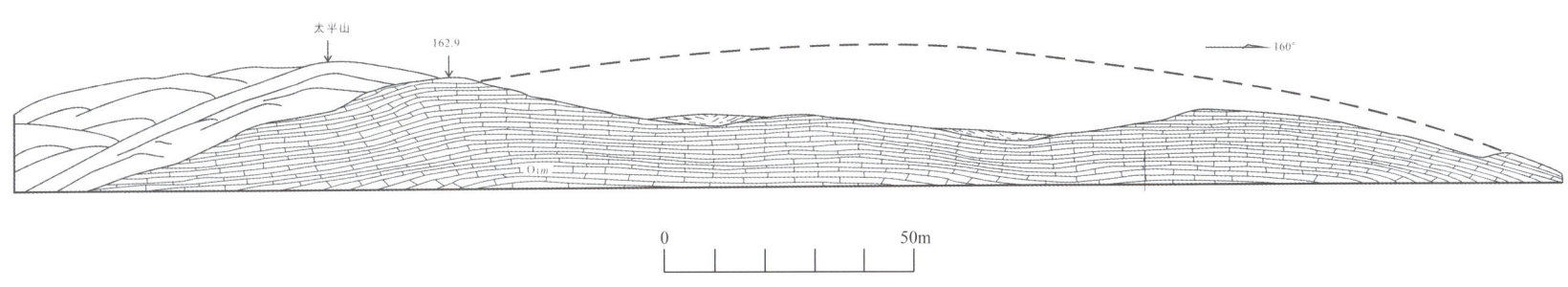

164 背斜构造剖面示意图（秦松贤编制）
Cross-section sketch map of the highpoint 164 anticline (complied by Qin Songxian)

孤山口复杂褶皱
The compound fold in Gushankou

    孤山口复杂褶皱地处区域性三岔背斜向北东倾伏端的北翼，组成地层为中元古界雾迷山组(Pt_2w)白云岩。由于所处特殊的构造部位，剪应变强烈，因而褶皱主体形态以斜歪倒转的"Z"形为主，并以次级褶皱大小不一、形态多样、伴派生小构造发育、构造变形复杂为特征。

    The Gushankou compound fold regionally occurs as a northern limb part of the "Sancha" anticline plunging to the northeast, which is constructed by dolomites of the Mesoproterozoic Wumishan Formation (Pt_2w). As a result of its special structural location that yielded intensive shear strain, the compound fold is characterized by inclined, overturned and Z-shaped secondary folds in various scales and configurations, and by many associated structures with complex deformation.

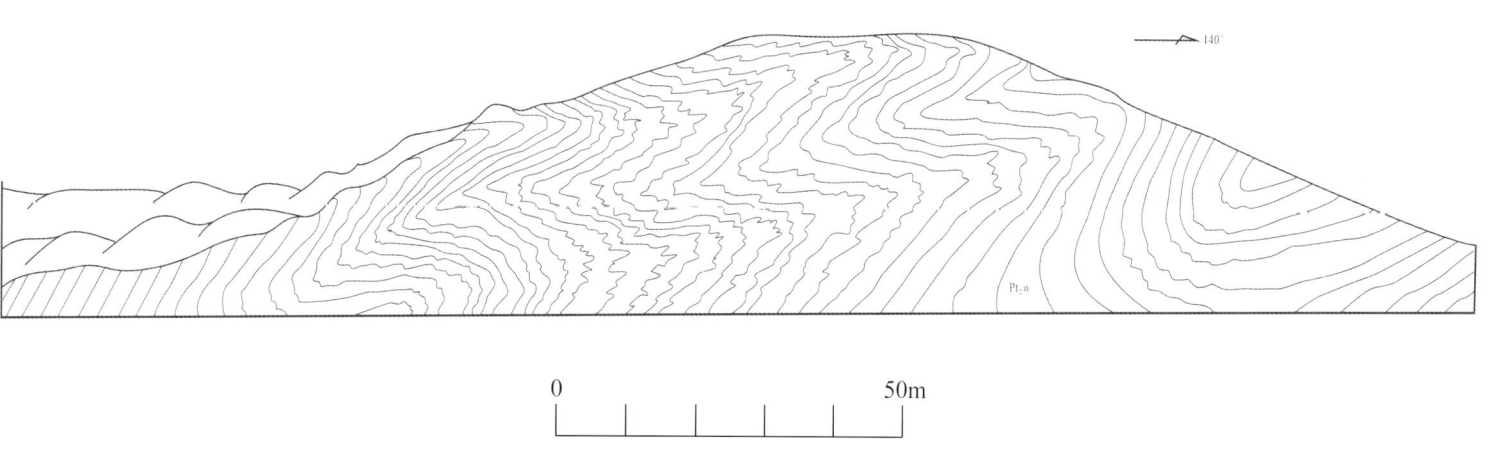

孤山口复杂褶皱构造剖面示意图(秦松贤编制)
Cross-section sketch map of the compound fold in Gushankou (Compiled by Qin Songxian)

次级褶皱呈"W"或"M"型组合
M-and W-shaped secondary folds (Photographed in Gushankou in Zhoukoudian)

右图为孤山口复杂褶皱中的低级别褶皱组合，形态呈"W"或"M"型，从而可判定其所处位置在高一级褶皱的内侧近转折端处。（摄于周口店弧山口）

The right figure shows M-and W-shaped secondary folds of the compound fold in Gushankou. Secondary fold patten indicate that the M-and W-shaped folds are located at the hinge zone of primary fold.

次级斜歪倾伏型褶皱
Inclined and plunging secondary folds

裸露良好的次级斜歪倾伏型褶皱，多项褶皱要素可以识别，如核部、转折端、两翼、枢纽、层面擦痕、轴面及轴面劈理等。（摄于周口店孤山口）

The inclined and plunging secondary fold elements, including core, hinge zone, limbs, hinge line, axial plane, axial cleavages and slickensides on fold surface, all are vivid in recognition and measurement. (Photographed in Gushankou in Zhoukoudian)

相似褶皱核部的变形三角区
Triangular deformation pattern in the core of similar folds

中元古界洪水庄组地层中出露的相似褶皱核部的变形三角区（铅笔指示处及左侧红褐色区）。（摄于周口店八角寨）
The triangular deformation pattern (indicated by a pencil and rufous area at the left) occurs in the core of similar folds constructed by the Hongshuizhuang Formation of Mesoproterozoic. (Photographed in Bajiaozhai in Zhoukoudian)

斜歪倒转背斜
Overturned and inclined anticline

复杂褶皱局部由多级褶皱组合的斜歪倒转背斜及其伴生小构造构成。（摄于周口店孤山口）
The complex folds consist of an overturned and inclined multi-compound anticline that is accompanied by associated structures. (Photographed in Gushankou in Zhoukoudian)

小型复式背斜
Minor compound anticline

由多个次级宽缓背斜与紧闭向斜组成的小型复式背斜。（摄于周口店孤山口）

The minor compound anticline consists of several secondary gentle anticlines and tight synclines (Photographed in Gushankou in Zhoukoudian)

上寒武统泥质条带灰岩中发育的斜歪倒转褶皱及轴面劈理与层理的置换关系（摄于周口店黄院东山梁）

The inclined and overturned fold in banded pelitic limestones of Upper Cambrian, where beddings are displaced by axial surface cleavages (Photographed in the eastern ridge of Huangyuan)

鞘褶皱(YZ 面)
Sheath fold(YZ plane)

鞘褶皱在横剖面上(YZ 面),因其封闭呈圆形、扁圆形或豆荚状,形似刀鞘而得名鞘褶皱。(摄于周口店孤山口)

On the cross section (YZ plane), sheath folds occur as closed circles, ellipses and/or legume shapes similar to a sheath, according to which the structure is named. (Photographed in Gushankou in Zhoukoudian)

鞘褶皱(XY 面)
Sheath fold(XY plane)

在平面上(XY面),鞘褶皱主要表现为舌状,且面上有大量的 A 型拉伸线理发育(尺子所示方向)。(摄于周口店孤山口)

On the horizontal section (XY plane), sheath folds occur as tongue shapes, where a number of A-typed extensional lineations also appear (indicated by rule strike). (Photographed in Gushankou in Zhoukoudian)

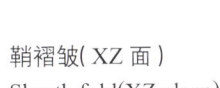

鞘褶皱(XZ 面)
Sheath fold(XZ plane)

在纵剖面上(XZ面),鞘褶皱则表现为 Z 形或 S 形,可借以判定韧性剪切带的相对剪切运动方向。(摄于周口店孤山口)

On the longitudinal section (XZ plane), sheath folds occur as Z-and/or S-shaped folds, which indicate the motion of ductile shear zone. (Photographed in Gushankou in Zhoukoudian)

复杂褶皱中的层间豆夹状 A 型褶皱（摄于周口店孤山口）
A legume-shaped A-typed fold developed in a compound fold (Photographed in Gushankou in Zhoukoudian)

第四部分 构造地质

S 形顺层平卧褶皱
S-shaped recumbent fold

洪水庄组千枚状板岩中发育的 S 形顺层平卧褶皱（褶叠层）。（摄于周口店八角寨）
The S-shaped recumbent fold occurs in phyllitic slates of the Hongshuizhuang Formation. (Photographed in Bajiaozhai in Zhoukoudian)

石香肠化平卧紧闭褶皱
Recumbent tight and boudinaged fold

石香肠化平卧紧闭褶皱，采自 164 背斜核部下奥陶统马家沟灰岩。（摄于周口店实习站标本）
The recumbent tight and boudinaged fold is developed in limestones of the Majiagou Formation of Lower Ordovician at the core of the highpoint 164 anticline. (Photographed from samples in the Base)

宽缓褶皱
Gentle folds

上寒武统泥质条带灰岩中发育的宽缓褶皱。（摄于周口店山顶庙）
The gentle folds are developed in banded pelitic limestones of the Upper Cambrian. (Photographed at the Shandingmiao in Zhoukoudian)

不协调弯流褶皱
Flexure-flow disharmonic folds

上寒武统泥质条带灰岩中发育的不协调弯流褶皱。（摄于周口店山顶庙）
The flexure-flow disharmonic folds are developed in banded pelitic limestones of the Upper Cambrian. (Photographed at the Shandingmiao in Zhoukoudian)

膝折及膝折带

Kink and kink band

膝折及膝折带，是一种兼具弯滑褶皱作用和剪切褶皱作用两种特征的特殊褶皱作用。膝折中的滑动褶皱作用常集中发生在不对称膝折的短翼部分，形成剪切带，故称膝折带。（周口店实习站标本照）

Kink and kink band are special folds characterized by both flexural-slip folding and shear folding. A kink band occurs when flexural slipping concentrates at a short limb of the asymmetrical fold. (Photographed from samples in the Base)

膝折带构造
Kink band structure

泥质条带灰岩中发育的膝折带。(摄于周口店三不管沟)

The kink band occurs in banded pelitic limestones.(Photographed in Sanbuguangou in Zhoukoudian)

膝折现象
Kink band

白云岩中发育的流劈理膝折现象。(摄于周口店孤山口)

The kink band occurs in dolomites with flow cleavages. (Photographed in Gushankou in Zhoukoudian)

肠状褶皱
Ptygmatic folds

中元古界洪水庄组千枚状板岩中由燧石条带或石英脉形成的肠状褶皱。（摄于周口店八角寨）

The ptygmatic folds are indicated by deformed cherts and quartz veins in phyllitic slates of the Hongshuizhuang Formation of Mesoproterozoic. (Photographed in Bajiaozhai in Zhoukoudian)

平卧褶皱
Recumbent folds

中元古界薄层白云岩中发育的平卧褶皱。（摄于周口店李各庄附近）

The recumbent fold occurs in thin-bedded dolomites of the Mesoproterozoic. (Photographed near the Ligezhuang Village in Zhoukoudian)

4.5 断层
Faults

断层是地壳岩石中沿破裂面发生明显位移的一种断裂构造,发育广泛,规模大小不一,是地壳中最重要、最常见的一种构造类型。由于地壳深浅层次不同,断层的表现也不一样。浅层次条件下形成脆性断层,深层次条件下则形成韧性断层或称韧性剪切带。

Faults are cracks in crustal rocks along which evident slipping occurs. They are one of the most important and common structures in earth crust, which develop widely in various scales. Faults vary in features according to the depths they occur in. Brittle faults arise in shallow crust, while in deep curst they transform into ductile fault that also referred as ductile shear zone.

4.5.1 脆性断层
Brittle Faults

小型正断层
Minor normal fault

雾迷山组白云岩中发育的小断层。断面较平整光滑,倾角中等,上下盘岩层牵引现象清楚,属小型正断层性质。(摄于周口店孤山口)

The minor normal fault occurs in the dolomites of the Wumishan Formation, which has a slippery fault surface and a moderate dip angle. The fault shows evident drag folds in both hanging wall and footwall, which indicate a normal motion. (Photographed in Gushankou in Zhoukoudian)

小型逆断层
Minor reverse fault

八角寨雾迷山组白云岩中为石英脉充填的小断层,切割了层理及顺层分布的燧石条带。断面倾角较缓,石英脉充填后又石香肠化。据上、下盘层理的牵引现象,可判定其为逆断层性质。(摄于周口店八角寨)

The minor reverse fault, which is indicated by an intruded quartz vein intersecting beddings and cherts, occurs in dolomites of the Wumishan Formation in Bajiaozhai. The fault has a gentle dip angle and the quartz vein within fault trace is boudinaged. The drag folds in both hanging wall and footwall suggest a reverse motion on the fault surface. (Photographed in Bajiaozhai in Zhoukoudian)

平移断层
Strike slip fault

房山花岗岩体西部边缘发育的一组东西走向的平移断层,岩体中的暗色包体被左行水平错移达120厘米(左图),和发育在主干断层旁侧的分枝小断层及其末端的马尾状小断层或节理(下图)。(摄于周口店龙门口)

The strike slip fault striking east-west is located at the western edge of the Fangshan granitic pluton, which gives a 120 cm sinistral slip to the dark intrusion (the figure left), and yields some branch to the main fault that transform into horse-tail faults at its terminal (the figure below). (Photographed at Longmenkou in Zhoukoudian)

马尾状小断层
Horsetail faults

房山花岗岩中东西向平移断层的东部,可见发育在主干断层尾端的马尾状小断层或节理。说明主干断层到此已趋于尖灭。(摄于周口店龙门口东山梁)

The horsetail faults occur at the terminal of the major fault, which is located to the east of the east-west strike slip fault in the Fangshan granitic pluton. (Photographed in the eastern ridge of Longmenkou in Zhoukoudian)

断层面上发育的擦痕和阶步（摄于周口店太平山南坡采石场）

Fault slickensides and steps (Photographed at the quarry in southern slope of the Taiping Mountain in Zhoukoudian)

断裂带内发育的挤压构造片岩（摄于周口店羊屎沟）

Compressional schists developed in a fault zone (Photographed at Yangshigou in Zhoukoudian)

断层的断坪、断坡
Fault flat and ramp

雾迷山组白云岩中发育的断坪（顺层）、断坡（切层）组合的断层。（摄于十渡景区八渡）
The fault flat (along bedding) and ramp (intersect beddings) occur in dolomites of the Wumishan Formation. (Photographed at Badu in the Shidu Scenic Spot area)

三不管沟断层破碎带
Fault damaged zone

呈东西走向的断层破碎带（剖面）。断层产状较陡，断层面总体南倾（右），上盘为上寒武统泥条灰岩，下盘为中寒武统粉砂质板岩（已出图），破碎带主要由红色断层泥、构造千枚岩及侵入的闪长岩脉（风化后呈灰白色）等组成。（摄于周口店太平山北坡三不管沟）

The section shows an east-west striking fault damaged zone, which has a steep fault surface dipping to the south (the right). The hanging wall consists of banded pelitic limestones of Upper Cambrian, and footwall of silt sand slates of Middle Cambrian, while the fault damaged zone mainly consists of red fault gouge, structural phyllites and intruded diorite veins (weathered to be gray). (Photographed at Sanbuguangou in north slope of the Taiping mountain in Zhoukoudian)

一条龙头处断层破碎带

Fault damaged zone

位于一条龙头处，发育的呈东西向产出的断层破碎带。断层产状较陡，总体南倾（右），上盘为新元古界下马岭组千枚状板岩，下盘为中元古界铁岭组白云质大理岩，带内由构造碎裂岩、断层泥及侵入的已蚀变花岗岩脉等组成。晚期性质应为正断层。（摄于周口店东山口）

The fault damaged zone, which is located at the head of Yitiaolong, strikes east-west and dips to the south (the right) steeply. The hanging wall consists of phyllitic slates of the Neoproterozoic Xiamaling Formation, and footwall of dolomitic marbles of Mesoproterozoic Tieling Formation, while the fault-damaged zone consists of structural breccia, fault gouge and altered intruded granitic veins. Normal motion in later stage is observed on the fault surface. (Photographed in Dongshankou in Zhoukoudian)

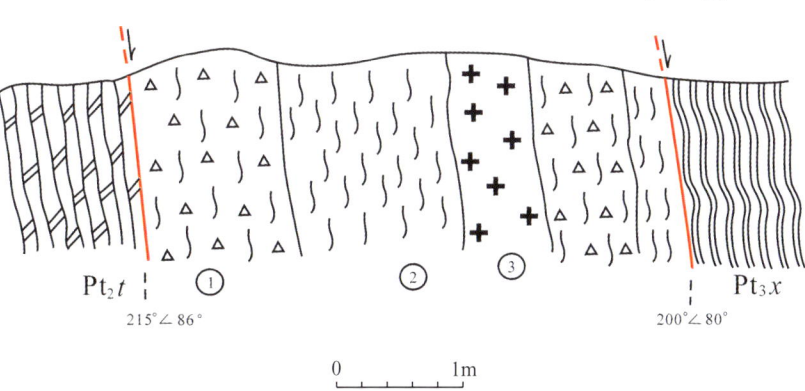

一条龙头处断层破碎带构造剖面图（秦松贤编制）
Fault structure sketch map of section across Yitiaolong (Compiled by Qin Songxian)

Pt_3x 下马岭组千枚状板岩。
Pt_2t 铁岭组白云质大理岩。
①构造碎裂岩；②断层泥；③蚀变花岗岩脉。
Pt_3x phyllitic slates of the Xiamaling Formation.
Pt_2t dolomitic marbles of the Tieling Formation.
①structural breccia; ②fault gouge; ③altered granitic vein.

房山西断层构造带（房山西断层）
West Fangshan Fault zone (West Fangshan Fault)

　　房山西断层，在区域上属于八宝山－南大寨断层带的组成部分，为周口店地区规模最大的断层。其上盘为中新元古界，下盘为早古生界马家沟组灰岩，中间缺失大套地层。断层带宽约 50 余米，带内结构复杂，但构造分带清楚、多期构造活动特征明显，属早期强烈挤压逆冲断层性质、晚期正断层性质。其形成时代为燕山期，但目前仍在活动。（摄于周口店房山西）

　　West Fangshan Fault, which is the largest fault in the Zhoukoudian area, consists a segment of the regional Babaoshan-Nandazhai Fault. The hanging wall consists of Mesoproterozoic and Neoproterozoic while the footwall consists of limestones of the Majiagou Formation of the Lower Paleozoic, between which massive strata loss occurred. The fault zone about 50m wide bears a complex internal structure, in which, however, structural division and several tectonic events are recognizable. The fault experienced an early intense compressional thrusting and late extension by normal faulting, which initiated in the Yanshanian period and still have present activation. (Photographed at west of Fangshan in Zhoukoudian)

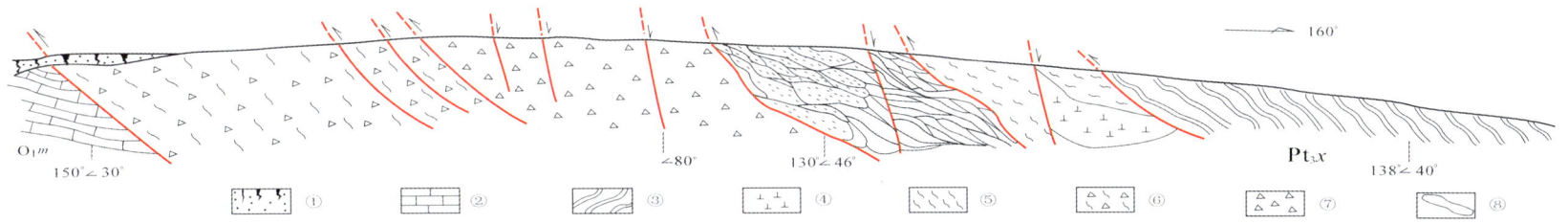

房山西断层构造带剖面图（秦松贤编制）
Fault structure sketch map of the section across west of Fangshan (Compiled by Qin Songxian)

①第四系沉积物；②下奥陶统马家沟灰岩；③新元古界下马岭组千枚状板岩；④碎裂蚀变闪长岩脉；⑤新元古界龙山组上段板岩挤压片理化带；⑥新元古界龙山组上段板岩挤压揉皱带；⑦新元古界龙山组下段变质砂岩碎裂角砾岩带；⑧新元古界龙山组下段变质砂岩挤压透镜体带。

①Quaternary;②limestones of the Lower Ordovician Majiagou Formation;③phyllitic slates of the Neoproterozoic Xiamaling Formation;④altered diorite veins;⑤compressional schistositized zone of phyllitic slates of the upper Member of Neoproterozoic Longshan Formation;⑥structural zone of slates of the upper Member of the Neoproterozoic Longshan Formation;⑦fault breccia of metamorphic sandstones of the lower Member of the Neoproterozoic Longshan Formation;⑧compressional tectonic lens zone of the lower Member of the Neoproterozoic Longshan Formation.

构造透镜体（一）
Tectonic lens (1)

由变质石英砂岩组成的挤压构造透镜体带和早期缓倾断面（左侧阴影处）被晚期陡倾断面（右侧及阴影处）所切割。（摄于房山西断层带局部）

The tectonic lens zone, which consists of metamorphic sandstones, in which late steep faulting (at the shadow right) cuts early gentle fault surfaces (at the shadow left). (Photographed at local of the West Fangshan Fault in Zhoukoudian)

构造透镜体
Tectonic lens

由变质石英砂岩构成的不同大小的构造透镜体。（摄于周口店房山西断层局部）

The tectonic lenses consist of metamorphic quartz sandstone. (Photographed at local of the West Fangshan Fault in Zhoukoudian)

挤压片理化带（摄于周口店房山西断层局部）

Compressional schistositized zone (Photographed at local of the West Fangshan Fault in Zhoukoudian)

沿断层带侵入的已蚀变破碎的闪长岩脉（锤把指处）（摄于周口店房山西断层局部）

Diorite veins with alteration and fragmentation that are intruded along the fault zone at the hammer. (Photographed in local of the West Fangshan Fault in Zhoukoudian)

房山十渡构造剖面
Structural section in Shidu in Fangshan

中元古界雾迷山组水平岩层中发育的断层褶皱破碎带。图中可见其左侧小部分和右侧大部分为水平岩层，图面靠左侧岩层产状变化明显，与两侧截然不同。（摄于房山十渡景区一渡）

The fault and fold zone is developed at the horizontal strata of the Mesoproterozoic Wumishan Formation. It's indicated that minor strata at left and most at right are horizontal; while the strata between change dramatic in attitude. (Photographed in Yidu of the Shidu Scenic Spot in Fangshan)

4.5.2 韧性断层（韧性剪切带）
Ductile fault (Ductile shear zone)

韧性断层是岩石在塑性状态下发生连续变形的狭窄高剪切应变带。不出现破裂或不连续面，具有"断而未破，错而似连"的变形特点。

图为车厂花岗岩中发育的一组走向北东的小型韧性剪切带，暗色包体被韧性错断。由此可判定其剪切运动方向为右行。（秦松贤摄于周口店车厂）

Ductile faults are narrow high shear strain zones occurring as continuous deformation in rocks under plastic state, in which no cracks or discontinuity in deformation occur. The figure shows a set of minor ductile shear zones striking northeast that offset a dark inclusion in granite at Chechang, which indicate a dextral shearing motion. (Photographed in Chechang in Zhoukoudian)

小型韧性剪切带
Minor ductile shear zone

房山花岗岩体中发育的一组走向北西的小型韧性剪切带。暗色包体被韧性剪切而形成曲颈瓶构造，由此可判定其剪切运动方向为左行。（摄于周口店车厂）

A set of minor ductile shear zones striking northwest occur in the granite of the Fangshan, where a dark intrusion sheared indicates a sinistral motion. (Photographed by Qin Songxian in Chechang in Zhoukoudian)

眼球状旋转残斑
Rotational mortar augen structure

顺层韧性剪切带内已大理岩化的白云岩岩块形成的眼球状旋转残斑显示左行剪切。（摄于周口店孤山口）

The rotational mortar augen structure is indicated by marbleized dolomite that is deformed by bedding ductile sinistal shearing. (Photographed in Gushankou in Zhoukoudian)

花岗质糜棱岩
Granitic mylonite

韧性剪切带内的花岗质糜棱岩。（摄于周口店官地村南东）

The granitic mylonite occurs within a ductile shearing zone. (Photographed at the southeast of the Guandi Village in Zhoukoudian)

糜棱岩的 S—C 组构
S-C fabric of mylonite

　　房山复式花岗岩西部边缘韧性剪切带中糜棱岩的 S-C 组构显示左行剪切。（平行笔杆的为 C 面理，与其斜交的为 S 面理）。（摄于周口店车厂）

　　The S-C fabric of mylonite occurs within a ductile shearing zone at the west of Fangshan Composite granite. (The C foliation parallels the pen, and the S foliation crosses the pen).(Photographed in Chechang in Zhoukoudian)

5

第四纪（含上新世）地层及地貌
Quaternary (including the Pliocene) and Geomorphology

　　周口店地区第四纪（含上新世）地层分布广泛，层序清楚，成因类型多样，生物化石丰富，其中，北京第一地点剖面最为著名，是华北地区中更新世的标准剖面，对我国的第四纪地层的划分和对比有着重要意义。

　　周口店地区位于太行山北端的山地与华北平原的过渡地带，属于低山丘陵地貌。地形总的趋势是从西北向东南逐渐降低。西北部峰岭峻峭，谷深坡陡，为侵蚀中山；东南部是山前倾斜平原，地形平坦，一望无际；周口河由北西向南东流入华北平原。

　　Quaternary strata (including the Pliocene) are widely distributed in the Zhoukoudian area, with clear sequences and a variety of genetic types, and rich in fossil organisms. The stratigraphic section of the first site of Beijing is most famous, it is the standard section of the Pleistocene in North China and has great significance for the division and correlation of Quaternary strata.

　　Zhoukoudian area is located in the border area between the mountain area of north Taihang Mountain and the North China Plain. It belongs to low mountain and hill area. The general terrain is gradually lowering from the northwest to southeast. The middle mountain region formed by erosion in the northwest is featured by sheer mountain ridges, high peaks, deep valleys and steep slopes. The Piedmont clinoplain in the southeast is flat and vast. The Zhoukou River flows northwest to southeast, and finally into the North China Plain.

5.1 第四纪(含上新世)地层
Quaternary (including the Pliocene)

5.1.1 上新统(N₂)
Pliocene (N_2)

(1) 上新统岩溶洞穴地层
Karst strata of Pliocene

本区上新统主要分布在山前海拔约 150 米的夷平面上,在山区则残存于同期的宽谷内。

Pliocene strata in this area are mainly distributed on the planation surfaces at about 150m above sea level in piedmont regions, and some remaining in straths at the same period in mountain areas.

周口店"上砾石层"
"Upper Gravel Stratum" in Zhoukoudian

周口店"上砾石层",属鱼岭组上段。为岩溶洞穴堆积。分布高程 143.8 米,高出周口河 58 米。剖面可分为 8 层,主要为地下河、地下湖河和洞穴灰岩角砾岩组成。(摄于周口店龙骨山)

The whole "upper gravel stratum" in Zhoukoudian is of the upper Member of the Yuling Formation, and of karst accumulation type. It has an elevation of 143.8m, 58m higher than the Zhoukou River. The stratigraphic section can be divided into 8 layers, mainly composed of underground rivers, underground lakes, and cave limestone breccias. (The photo was taken at Dragon Bone Hill in Zhoukoudian)

湖相沉积
Lacustrine sediments

周口店"上砾石层"下部第①~②层。岩性为灰黄色砂质粉砂、钙质胶结粉砂和细砂。具薄的平行层理,由于后期变动,岩层发生倾斜变形。(摄于周口店龙骨山)

Lacustrine sediments have formed the first and second layers in the lower part of "Upper Gravel Stratum" in Zhoukoudian. The lithology include grey-yellow sandy silt, calcareous cemented silt and fine sand. The strata are characterized by thin parallel bedding. Owing to changes in late period, the strata have been distorted and tilted. (Photographed at Dragon Bone Hill in Zhoukoudian)

河流沉积物
Fluvial sediments

周口店"上砾石层"上部第⑥~⑦层。第⑥层为黄棕色钙质胶结黏土质粉砂层,具板状水平层理,厚1.4米。第⑦层岩性为灰白色钙质胶结的砂夹砾石透镜体,厚1.0米。(摄于周口店龙骨山)

Fluvial sediments have formed the sixth and seventh layers in the upper part of the "Upper Gravel Stratum" in Zhoukoudian. The lithology of the sixth layer is yellow-brown calcareous cemented clayey silt, with tabular horizontal bedding and a thickness of 1.4m. The lithology of the seventh layer is gray-white calcareous cemented gravel lens interbedded by sand, with a thickness of 1.0m. (Photographed at Dragon Bone Hill in Zhoukoudian)

顶盖堆积

Cap deposits

属周口店"上砾石层"上部第⑧层。岩性为灰白色钙质胶结的含灰岩角砾粗砂层，中间夹 0.2～0.5 米灰白色钙质胶结的细砂层。含哺乳动物化石，厚 3.2 米。（摄于周口店龙骨山）

Cap deposits formed the eighth layer in the upper part of the "Upper Gravel Stratum" in Zhoukoudian. The lithology of the eighth layer is gray-white calcareous cemented coarse sand with limestone breccia, interbedded by gray-white calcareous cemented fine sand which is 0.2~0.5m thick; this layer contains mammalian fossils and is 3.2m thick.(Photographed at Dragon Bone Hill in Zhoukoudian)

（2）上新统地表堆积
Pliocene surface sediments

上新统红土层
Pliocene red soil

上新统红土。位于龙骨山顶部，为上新世岩溶作用的风化溶蚀残余堆积。

Pliocene red soil is distributed at the top of Dragon Bone Hill. It generated from Pliocene karst corrosion and weathering.

5.1.2 下更新统（Q_1）
Lower Pleistocene（Q_1）

下更新统太平山组剖面
Section of the Taipingshan Formation of Lower Pleistocene

下更新统太平山组（Q_1t）。位于太平山北坡大砾岩山下，海拔高程 140 米。剖面总厚约 15 米，共有 13 个自然岩性层组成。顶部约 4 米为周口店组（Q_2zh）。

The Taiping Mountain Formation of Lower Pleistocene (Q_1t) is distributed at nortnern slope of the Taiping Mountain and the foot of the Daliyan Mountain, with an elevation of 140m. The section is about 15m thick, and consists of 13 natural lithological layers. The top about 4m is of the Zhoukoudian Formation (Q_2zh).

5.1.3 中更新统（Q_2）
Middle Pleistocene (Q_2)

周口店龙骨山第一地点西壁剖面
The Xibi Section at Spot 1 in the Dragon Bone Hill, Zhoukoudian

周口店组（Q_2zh）——周口店第一地点西壁剖面。位于龙骨山东北坡，剖面最高点海拔128米，剖面总厚度达30米。剖面共有13层组成。其岩性总体上为弱钙质胶结的石灰岩角砾组成，其中第⑩层为杂色灰烬层，厚约0.65m。

The Xibi Section at Spot 1 in the Dragon Bone Hill of Zhoukoudian is of the Zhoukoudian Formation (Q_2zh). It is located at the northeastern slope of Dragon Bone Hill, with a maximum elevation of 128m and a total thickness of 30m. The section consists of 13 layers. The lithology are mainly weak calcareous cemented limestone breccia, while the tenth layer is a varicolored ash layer with a thickness of about 0.65m.

5.1.4 上更新统（Q₃）
Upper Pleistocene (Q_3)

（1）洞穴堆积
Cave deposits

溶洞灰岩角砾堆积物
Karst cave limestone breccia deposits

山顶洞组。位于周口店龙骨山山顶，堆积物为比较厚的灰岩角砾，夹 5 个文化层，角砾间充填灰黄色粉砂土，未胶结。全部堆积厚度达 10 米以上，含晚期智人(山顶洞人)化石及含大量哺乳动物化石等 48 种。年代学测定表明，堆积物形成于 10.2kaBP（^{14}C）～ 4.9kaBP(TL)。

The deposits are of the Shandingdong Formation, located at the top of Dragon Bone Hill in Zhoukoudian. They are thick limestone breccias, interbeded with 5 cultural layers. There are grey-yellow silt sands among the breccias. The layers are uncemented, with a total thickness of over 10m. The deposits include fossils of late Homo sapiens (Upper Cave Man) and 48 kinds of mammal fossils. Dating indicates that deposits were formed at 10.2 kaBP(^{14}C)~4.9 kaBP(TL).

（2）地表堆积
Surface sediments

黄土状土层剖面
Section of loess-like soil layer

黄土状土层。位于老牛沟沟头西侧。岩性为红黄色亚砂土，垂直节理发育，具大孔隙结构，土质较坚硬。上更新统黄土状土在区内沟谷广泛分布，如三不管沟、羊屎沟等。

Loess-like soil layer is located at the west of Laoniu Gully head. The lithology is red-yellow sandy loam, with vertical joints and macropore structure. The soil is hard. Loess-like soil of upper Pleistocene is widely distributed in the valley region of the area, such as Sanbuguan Gully and Yangshi Gully.

5.2 地貌
Geomorphology

构造-剥蚀中低山地貌

Low-medium mountain formed by tectonic denudation

周口店地区最高峰——上寺岭（猫耳山），海拔高度1 307米，其顶部为北台期夷平面。

The highest peak in the Zhoukoudian area is the Shangsiling (Maoer Mountain) at 1 037m, the mountaintop surface of which is Peitai stage planation surface.

岩溶-构造剥蚀中低山地貌
Low-medium mountain formed by karst-tectonic denudation

由石灰岩组成的岩石经溶蚀和流水侵蚀共同作用形成中低山。（黄院一带）
Rocks consisting of limestones formed low-medium mountain under the combined action of corrosion and moving water erosion. (Huangyuan area)

岩溶-构造剥蚀丘陵
Hills formed by karst-tectonic denudation

由石灰岩、碎屑岩组成，经岩溶-构造剥蚀作用形成的丘陵地貌景观（龙骨山－升平山－太平山）。由太平山向东为山前剥夷面。（摄于周口店镇周口河西岸）
Hilly landscape (Dragon Bone Hil-Shengping Mountain-Taiping Mountain) was formed under the combined action of karst and tectonic denudation and consisted of limestone and clastic rocks. The surface of the area to the east of the Taiping Mountain is piedmont planation surface of denudation. (Photographed at the west bank of Zhoukoudian Town.)

花岗岩剥蚀丘陵
Denudation hill of granite

由花岗岩岩石（房山岩体）组成，经风化、风蚀和流水侵蚀共同作用而形成的剥蚀丘陵地貌景观。（摄于周口店官地村北）
Denudation hill (Fangshan Pluton) comprised of granites was formed under the combined action of weathering, wind erosion and moving water erosion. (Photographed north in Zhoukoudian.)

房山岩体花岗岩剥蚀丘陵地貌景观（摄于周口店车厂村北）
Landscape of denudation hill of granite in the Fangshan Pluton (Photographed to the north of Chechang Village in Zhoukoudian)

河流地貌

Fluvial landform

拴马庄河河谷地貌。共有四级河流阶地发育。(摄于周口店拴马庄)
River-valley landform of the Shuanmazhuang River consists of the four river terraces.
(Photographed at Shuanmazhuang in Zhoukoudian)

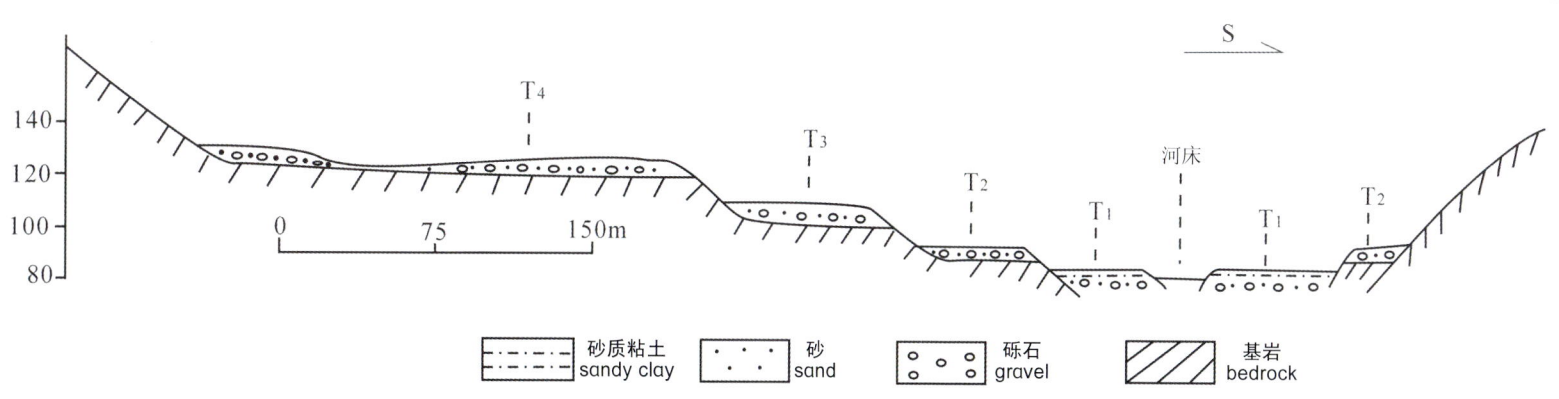

拴马庄河河谷阶地横剖面图
Cross section of the valley terrace of the Shuanmazhuang River

岩溶地貌——石芽
Karst landform-Clints

石灰岩表面残差不齐的小型突起。它是地表水沿节理裂隙进行溶蚀和侵蚀共同作用而形成的。(摄于十渡)

Clints are small jagged protrusions on the surface of limestone. They are formed under the combined action of corrosion and erosion caused by surface water which flowed through joint fissures. (Photographed at Shidu)

岩溶地貌——峰丛地貌
Karst landform-Peak cluster landform

由石灰岩组成的基座相连的峰林,峰与峰之间常成马鞍形谷地,是峰林形成的早期阶段,是溶蚀和侵蚀共同作用的产物。(摄于十渡风景区)

Peak cluster is a group of peaks with a common limestone basement, areas between peaks are always developed into valleys shaped like saddles. It's the early stage of the formation of peak forest, formed under the combined action of corrosion and erosion. (Photographed at Shidu Scenic spot)

岩溶地貌——石笋
Karst landform-Paleo-stalagmite

周口店龙骨山山顶洞中的古石笋。（摄于周口店龙骨山）

This is a paleo-stalagmite in the Upper Cave of Dragon Bone Hill in Zhoukoudian. (Photographed at Dragon Bone Hill in Zhoukoudian)

岩溶地貌——小型溶洞及溶孔
Karst landform-Minor karst cave and cavern porosity

地下水在沿岩石层面流动过程中溶蚀和侵蚀共同作用的产物。（摄于周口店太平山南坡采石场）

Formed under the combined action of corrosion and erosion caused by ground water which flowed through rock surface (Photographed the quarry on the south slope of Taiping Mountain in Zhoukoudian)

岩溶地貌——钟乳石
Karst landform-Stalactites

自溶洞顶部向下生长的一种以碳酸钙为主的沉积体。开始只成为一小突起附在洞顶，以后逐渐增长而成。（摄于房山区石花洞）

Stalactite is a deposit projecting downward from the roof of a karst cave, consisting mainly of calcium carbonate. It begins with a small projection attached on the cave ceiling, and then gradually grows up. (Photographed at Shihua Cave in Fangshan District)

岩溶地貌——石花
Karst landform-Stone flowers

呈丛花状散布在洞壁或其他洞穴堆积物表面的雾滴水沉积。亦可由因气温、湿度变化，产生密集的成簇状的凝结水珠所析出的碳酸钙沉积形成。（摄于房山区石花洞）

Stone flowers are flower-like clusters of deposits attached on the cave walls or on the surface of other cave deposits, formed by fog drip. Some are dense clusters of calcium carbonate which precipitated out of condensed water due to the variation of temperature or humidity.(Photographed at Shihua Cave in Fangshan District)

岩溶地貌——现代石笋
Karst landform-Modern stalagmites

溶洞内洞顶的水滴落到底板后，形成由下而上增长的碳酸钙沉积，因形如笋状而得名。（摄于房山区石花洞）

Modern stalagmite is a cylindrical mass of calcium carbonate projecting upwards from the floor of a karst cave, formed by precipitation from continually dripping water. They are called "Shisun" in Chinese because they are shaped like bamboo shoot. (Photographed at Shihua Cave in Fangshan District)

岩溶地貌——石柱
Karst landform-Stalacto-stalagmites

洞穴顶部石钟乳往下长，与之对应的洞穴底板的石笋向上长，两者生长相连接后所形成的柱状体。（摄于房山区石花洞）

Stalacto-stalagmite is a pillar in a cave, formed by a stalactite extending downward from the roof of the cave to meet with a complementary stalagmite extending upward from the floor of the cave. (Photographed at Shihua Cave in Fangshan District)

风蚀地貌——石蘑菇群
Deflation landscape-Mushroom rock groups

位于官地村北,房山花岗岩岩体由于节理的存在,沿节理软弱带侵蚀较强而形成细颈和断开,经风化和风蚀作用而形成石蘑菇状的地貌。有时强风可将蘑菇石上部吹得晃动,故亦称摇摆石。

The stones are located to the north of Guandi Village. Fangshan granitic pluton was formed into mushroom rock landscape under the action of erosion and weathering when the weak parts of the joint turned into narrow neck or fault with strong erosion. Sometimes the upper part of the mushroom rock can be shaken by strong wind, so it is also known as rocking stone.

5.3 主要不同成因的沉积物
Major accumulations of different genesis

5.3.1 残积物
Eluvium

碎屑型残积物
Detrital eluvium

碎屑型残积物是房山花岗岩经长期的物理风化作用而形成的。主要表现是母岩经风化破坏，岩石矿物的化学成分改变不大。（摄于官地村）

Detrital eluvium is formed by Fangshan granitic pluton as a result of a long-term physical weathering, mainly shown as: the parent rock was destroyed by weathering while the chemical constituents of rocks and minerals were rarely changed. (Photographed at Guandi Village)

5.3.2 坡积物
Colluvium

由亚砂土夹角砾层和角砾透镜体组成的坡积物。(摄于周口店拴马庄东)
The colluvium is composed of breccia lenses and sandy loam layer interbedded by breccia. (Photographed at the east of Shuanmazhuang of Zhoukoudian)

5.3.3 洪积物
Diluvial accumulation

洪积扇扇根沉积
Fanhead accumulation of a diluvial fan

洪积扇扇根沉积。由砾石层夹亚砂土透镜体构成的洪积物。砾石层有良好的叠瓦状沉积构造。砾石的最大扁平面倾向流水的上游，可用来指示水流的方向。（摄于太平山北坡羊屎沟）

The fanhead diluvial accumulation is composed of gravel layer interbedded by sandy loam lenses. The gravel layer has good imbricate sedimentary tectonics. The planes of maximum flat of gravels orient toward the upper reaches of the river, indicating the flow direction of the water. (Photographed Yangshi Gully at the northern slope of the Taiping Mountain)

洪积扇扇中沉积
Midfan accumulation of a diluvial fan

洪积扇扇中沉积。由砾石夹亚砂土构成的洪积物，砾石层有良好定向排列。（摄于太平山北坡羊屎沟）

The midfan diluvial accumulation is composed of gravels interbedded by sandy loam. The gravel layer has good orientation. (Photographed Yangshi Gully at the northern slope of the Taiping Mountain)

洪积扇扇缘沉积
Lower fan accumulation of a diluvial fan

洪积扇扇缘沉积。由亚砂土夹砾石构成的洪积物。（摄于太平山北坡羊屎沟）

Lower fan diluvial accumulation is composed of sandy loam interbedded by gravels. (Photographed at Yangshi Gully at the northern slope of the Taiping Mountain)

5.3.4 重力堆积物
Gravity accumulation

洞穴角砾层
Karst cave breccia layer

洞穴角砾层由溶洞崩塌堆积而成。（摄于龙骨山）

The karst cave breccia layer is formed by deposit from cave collapse. (Photographed at Dragon Bone Hill)

5.3.5 冲积物
Alluvium

具有二元结构的冲积物。（摄于周口店下中院）
The alluvium with binary structure.(Photographed in Xiazhongyuan in Zhoukoudian)

5.4 古人类与古动物
Palaeo-human and Palaeo animal

5.4.1 古人类
Palaeo-human

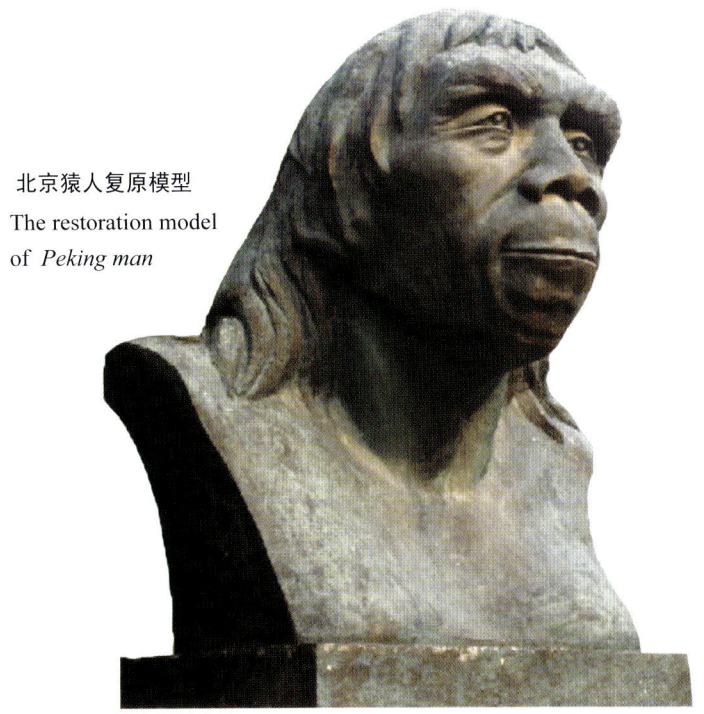

北京猿人复原模型
The restoration model of *Peking man*

(1) 古人类化石
Palaeo-human fossils

北京猿人

Sinanthropus pekinensis

北京猿人正式名称为"中国猿人北京种"。北京猿人生活在距今大约 70～20 万年。周口店北京猿人化石共出土头盖骨 6 具。北京猿人的颧骨较高，头部前倾，平均脑量达 1 088 毫升；据推算北京猿人身高为 156 厘米(男)，144 厘米(女)。北京猿人是属于从古猿进化到智人的中间环节的原始人类，其发现在生物学、历史学和人类发展史的研究上有着极其重要的价值。

Peking man, formally known as *Sinanthropus pekinensis*, lived about 700～200 thousand years ago. There were six *Peking man* skull fossils unearthed in Zhoukoudian of Beijing City. *Peking man* has high cheekbones, an average brain capacity of 1 088 milliliters, and the head is projected forward. It is estimated that *Peking man*'s height is about 156 cm for the male and 144 cm for the female. *Peking man* is the intermediate link between *Australopithecus* and *Homo erectus*. The discovery of this species is of great importance to the study of biology, history and human evolution.

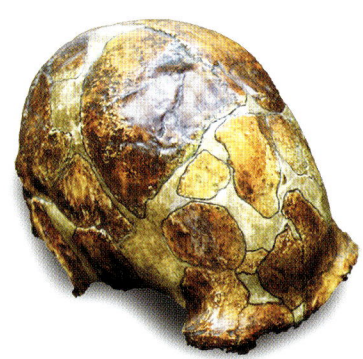

北京猿人第 X 号头盖骨(模型)
The Xth skull fossil of *Peking man* (model)

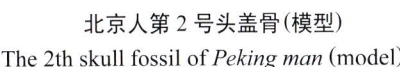

北京人第 2 号头盖骨(模型)
The 2th skull fossil of *Peking man* (model)

山顶洞人
Upper cave man

山顶洞人属发现于华北的晚期智人化石。头骨硕大，上面部低矮，整个面部中等程度的突出，眼眶较低，梨形孔较阔，其下缘呈鼻前窝形，下颌骨颏孔位置较低，且较靠后，颏部突度较小，接近现代黄种人。山顶洞人生活的时代约为晚更新世晚期。

山顶洞的人类化石共代表 8 个男女老少不同的个体。山顶洞人的体质已很进步，与现代人基本相一致。脑量已达 1 300～1 500 毫升。测年资料表明，山顶洞人生活于距今 2.7 万年至 3.4 万年前之间。山顶洞人仍用打制石器，但已掌握磨光和钻孔技术。他们已会人工取火，靠采集、狩猎为生，还会捕鱼。山顶洞人已用骨针缝制衣服，懂得爱美。他们死后还要埋葬。

Upper cave man was found in the late *Homo sapiens* fossils in Norts China. They have a large skull. The upper facial part is low and the whole face protrudes moderately. The orbit is also low. The piriform orifice is broad and the lower part forms a socket in front of the nose. The lower javo and mental foramen are relatively low and afterward . The chindoesn't protrude much does to modern yellow people. *Upper cave man* lived in the end of Late Pleistocene.

The human fossils found in the upper cave represent eight individuals of different ages and sexes. The constitution of *Upper cave man* is very progressive and as robust as that of Homo sapiens. The brain capacity was 1 300 to 1 500 milliliters. Dating showed that they lived between 34 to 27 thousand years ago. *Upper cave man* still used chipped stone tools, but they knew how to polish and drill holes, and how to fish and make fire by hand. Hunting and collecting were their major productive labor. They knew how to use bone needles to sew clothes and began to pursue beautiful things. After their death, they were buried.

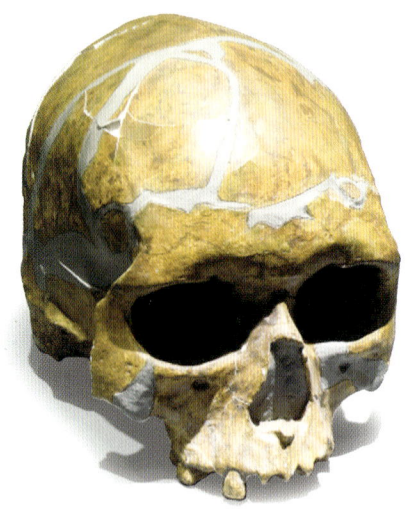

山顶洞人头骨 101 号
The 101th skull of *Upper cave man*

山顶洞人复原图
Restoration picture of *Upper cave man*

(2) 用火遗迹
Evidence for the use of fire

在北京猿人居住过的洞穴里，发现的灰烬层、烧石和烧骨等，表明北京猿人已懂得使用火、支配火、学会保存火种的方法，是人类由动物界跨入文明世界的重要标志。

Ash layers, burned stones and burned bones were found in the caves *Peking man* had lived in, indicating that *Peking man* had already known how to use fire, control fire and preserve the kindling. Use of fire is an important mark of civilization which separates man from other animals.

灰烬标本
Ash Sample

烧骨
Burned bones

灰烬层剖面
Ash Layers Section

烧石
Burned Stones

(3) 石器——北京猿人制作和使用的工具
Stone tools-Tools made and used by *Peking man*

石器是以岩石为原料制作的工具，它是人类最初的主要生产工具。北京猿人的石器以石片石器为主，石核石器较少，且多为小型。北京人的石器有砍砸器、刮削器、雕刻器、石锤和石砧等多种类型。原料有来自洞外河滩的脉石英、砂岩、石英岩、燧石等砾石，也有从两公里以外的花岗岩山坡上找来的。

Stone tools are tools made of stones. They were the first primary means of production in early times. *Peking man* mainly used flake stone tools and little core stone tools. The stone tools *Peking man* used were mostly small, including stone chopper, scraper, graver, hammer, anvil, and so on. The raw materials were vein quartz, sandstone, quartzite, chert, or other kinds of gravel, from the riverbank out of the cave, or from granite hillside two kilometers away.

尖状器
Cuspate stone tool

刮削器
scraper

砍砸器
stone chopper

石砚
stone inkstone

石斧
stone ax

（4）骨器
Bone tools

山顶洞人时期已经有了制作精美的磨制骨器。骨角器中最有代表性的是骨针，针身保存完好，仅针孔残缺，残长 82 毫米，针身浑圆微弯，针尖如芒，针孔是用小而细锐的尖状器挖成的。它是中国最早发现的旧石器时代的缝纫工具。骨针的出现意味着当时已会缝纫。在山顶洞人的洞穴里还发现了一些有孔的兽牙、骨管、海蚶壳和磨光的石珠，可能是他们佩戴的装饰品，表明山顶洞人已经有了审美观念。

Delicate ground bone tools had already existed in the period of *Upper cave man*. Bone needle was a representative of the bone & horn tools. The needle body is smooth, round, slightly bent and perfectly preserved, while the pinhole is sharp and has been damaged, was made by small and sharp objects. The incomplete needle is 82 millimeters long. It's the first sewing tool of the Paleolithic Age found in China. The advent of the bone needle means that the people of the age were able to sew clothes. Perforated animal teeth, bone tubes, sea clam shells and polished stone beads were also found in the Upper Cave and thought to be decorations which indicate that upper cave man had already got aesthetic perception.

骨针
bone needle

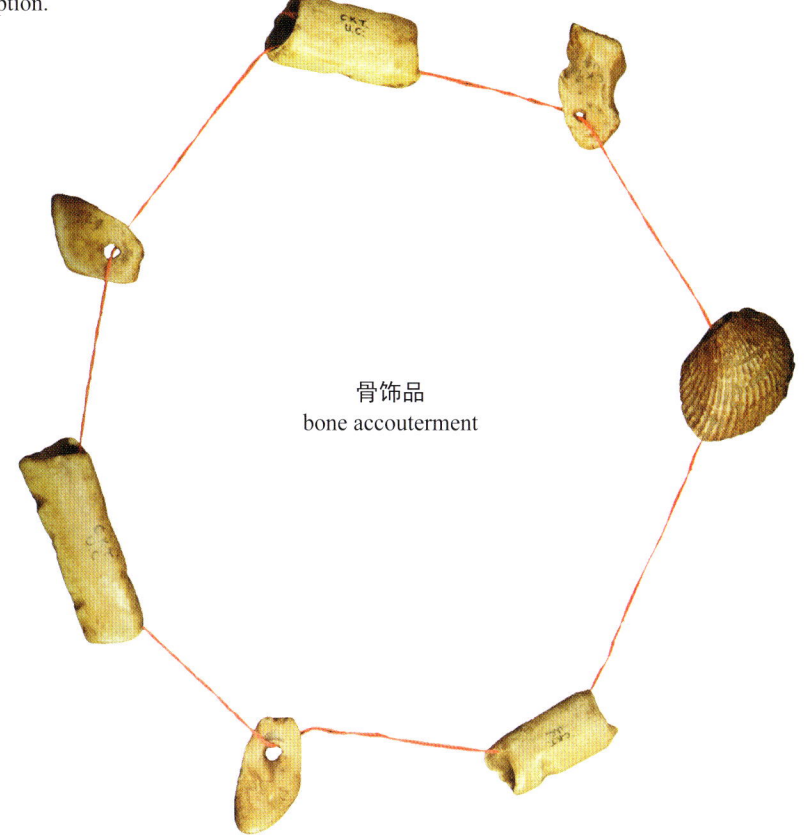

骨饰品
bone accouterment

5.4.2 古动物化石
Palaeo animal fossils

(1) 周口店动物群
Zhoukoudian fauna

周口店动物群（Zhoukoudian fauna）是华北地区更新世中期的一个哺乳动物群，它以北京周口店第一地点洞穴堆积中的化石群为代表，是与北京猿人同时期的一个动物群。化石种类相当丰富，仅周口店第一地点发现的哺乳类就有 94 种。其中有些是更新世早期动物群的残留种或其变种，如三门马（*Equus sanmeniensis*）、梅氏犀（*Rhinoceros mercki*）、纳玛象（*Palaeoloxodon namadicus*）、中国鬣狗（*Hyaena sinensis*）等。有些是更新世中、晚期所特有的新种，如肿骨鹿（*Sinomegaceros pachyosteus*）、葛氏斑鹿（*Pseudaxis grayi*）、洞熊（*Ursus spelaeus*）、最后斑鬣狗（*Crocuta ultima*）、最后剑齿虎（*Machairodus inexpectatus*）、杨氏虎（*Panthera youngi*）、德氏水牛（*Bubalus teilhardi*）等。

Zhoukoudian fauna is a mid-Pleistocene mammal fauna in North China. It is represented by the fossil group found in the cave deposits in the Zhoukoudian first site. It's at the same age as Peking man and rich in fossil species. There are 94 species of mammal just in Zhoukoudian first site. Some are residual species or altered species of early Pleistocene fauna, such as *Equus sanmeniensis*, *Rhinoceros mercki*, *Palaeoloxodon namadicus*, *Hyaena sinensis*, and so on. Some are new species peculiar to middle and late Pleistocene, such as *Sinomegaceros pachyosteus*, *Pseudaxis grayi*, *Ursus spelaeus*, *Crocuta ultima*, *Machairodus inexpectatus*, *Panthera youngi*, *Bubalus teilhardi*, and so on.

中国鬣狗骨架化石
The skeleton fossil of *Hyaena sinensis*

双角犀右上颌
The right upper jaw of *Biangular rhinoceros*

双角犀左下颌
The left lower jaw of *Biangular rhinoceros*

肿骨大角鹿
Megaloceros pachyosteus

葛氏斑鹿角
The born of *Pseudaxis grayi*

葛氏斑鹿角头骨
The horn skull bone of *Pseudaxis grayi*

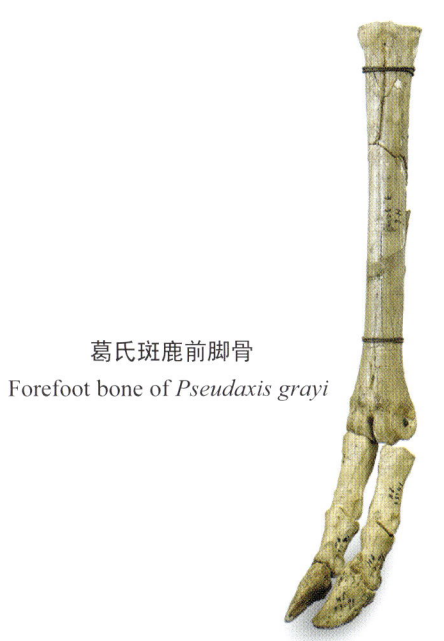

葛氏斑鹿前脚骨
Forefoot bone of *Pseudaxis grayi*

（2）山顶洞动物群
Shandingdong fauna

山顶洞动物群（Shandingdong fauna）是指中国华北地区更新世晚期偏晚时期的一个哺乳动物群，它以北京市周口店山顶洞洞穴堆积中的化石群为代表，是与山顶洞人同时期的一个动物群。包括洞熊（*Ursus spelaeus*）、最后斑鬣狗（*Crocuta ultima*）、虎（*Panthera tigris*）、豹（*Panthera pardus*）、猎豹（*Cynailurus*）、狼（*Canis lupus*）、狐（*Vulpes*）、豺（*Cuonalpinus*）、獾（*Meles*）、野驴（*Equus hemionus*）、斑鹿（*Pseudaxis*）、赤鹿（*Cervus canadensis*）、野猪（*Sus* sp.）、象（*Elephas*）、牛（*Bos*）、羊（*Ovis*）等哺乳动物。哺乳动物中除相当一部分现生种外，也有几种现代已绝灭的，如洞熊、最后斑鬣狗。

Shandingdong fauna is a mammal fauna of the later Late Pleistocene in North China. It is represented by the fossil group found in the cave deposits in Shandingdong of Zhoukoudian. It's at the same age as Upper Cave Man. It includes *Ursus spelaeus*, *Crocuta ultima*, *Panthera tigris*, *Panthera pardus*, *Cynailurus*, *Canis lupus*, *Vulpes*, *Cuonalpinus Meles*, *Equus hemionus*, *Pseudaxis*, *Cervus canadensis*, *Sus* sp., *Elephas*, *Bos*, *Ovis* and other mammal species. Except several species extinct, such as *Ursus spelaeus* and *Crocuta ultima*, quite a few of the mammal species are extant species.

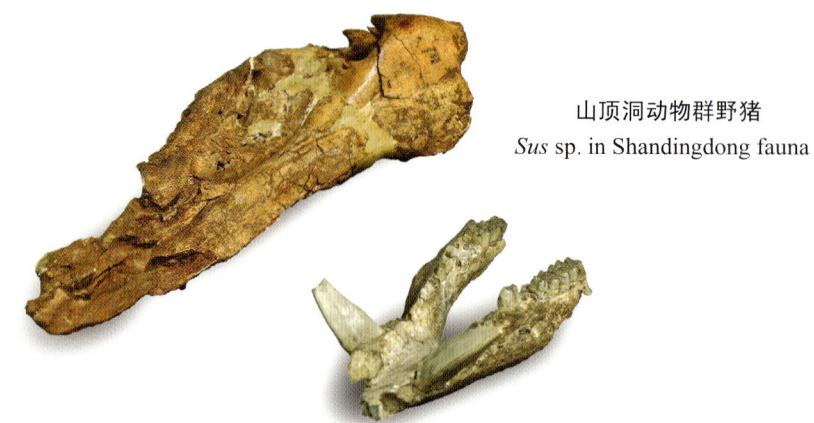

山顶洞动物群野猪
Sus sp. in Shandingdong fauna

意外巨剑齿虎（复原图）
M. inexpectatus(restoration picture)

意外巨剑齿虎的头骨及牙齿
The skull and tooth of *M. inexpectatus*

变异狼头骨
The skull of *Aberrance wolf*

6 区域资源与环境保护

Regional Resources and Environmental Protection

周口店地区的矿产资源比较丰富，主要是非金属矿产。有煤、石灰岩、白云岩、花岗岩、大理岩、红柱石、石墨、耐火粘土等十余个矿种。金属矿产除铁、铅锌矿点和矿化外，尚未发现有工业价值的矿床。

北京房山是我国各类地质现象比较完全和集中的地区之一，尤其是因其毗临首都、交通便利，占有自然地理和人文地理的优势，故在本区开发地质、人文旅游事业前景十分广阔。

保护和改善环境，协调人类与环境的关系，保障经济社会的持续发展是一项长期艰巨的任务，必须受到社会各方面的重视。

Zhoukoudian area is rich in mineral resources, primarily non-metallic minerals. There are more than ten kinds of non-metallic minerals, such as coal, limestone, dolomite, granite, marble, andalusite, graphite and refractory clay. While there are only two kinds of metallic mineral resources, Fe and Pb-Zn, rich enough to be commercially mined.

Fangshan in Beijing concentrates almost all the geological phenomena in China. Due to its adjacent location to the capital and convenient traffic, it has the advantages of both physical and human geography, so that the geological and cultural tourism of this district has a bright future.

Protecting and improving the environment to coordinate the relationship between humans and the environment and support the sustainable economic and social devolopment is a long-term arduous task ,must be paid attention to by all aspects of society.

6.1 矿产资源
Mineral Resources

6.1.1 煤炭
Coal

煤是周口店地区最重要的矿产资源。埋藏于北部凤凰山至上寺岭一带的工业用煤,已探明储量数亿吨。现建有房山煤矿、长沟峪煤矿等中型矿山多处,年采掘量达百万吨级。主要开采煤层为下侏罗统窑坡组。该组含可采煤4～7层。煤质均为无烟煤。

在太平山、升平山、凤凰山南麓、黄院北山等地分布的太原组和山西组地层中,含有2～4层凸镜体状或串珠状薄煤层,为民办小煤窑开采的对象。

Coal is the most important mineral resource of Zhoukoudian area. In the areas from northern Fenghuang Mountain to Shangsiling Mountain, the coal for industrial use was measured to be hundreds of million tons. There are many middle sized coal mines, such as Fangshan Colliery and Changgouyu Colliery, with an annual exploitable amount of megatons. The main coal seams are exploited from the Yaopo Formation of Lower Jurassic which consists of 4~7 beds of antnracites.

In the Taiyuan and Shanxi Formations which distribute in Taiping mountain, Shengping Mountain, Fenghuang Mountain south slope, north part of Huangyuan and some other places, the lens shaped and bead shaped thin coal seams are mainly exploited by small private coal mines.

周口店长沟峪煤矿
Changgouyu Colliery in Zhoukoudian

6.1.2 花岗石
Granite

周口店地区生产的花岗石石料颇负盛名,开采对象主要是房山复式岩体的边缘相和过渡相以及稍早侵入的石英闪长岩体,后者颗粒细而匀,属上等石料。宏伟的天安门广场就曾经采用了大量的房山花岗石。

The stock of granite is one of the most famous products in Zhoukoudian area. Some quarries are in border facies and transitional facies of the Fangshan multi-intruded pluton and some in early quartz diorite intrusions where granite is fine grained, uniform and first-class. A large number of Fangshan granite is used in Tiananmen Square.

周口店官地村花岗石采石场

Granite Quarry in Guandi Village Zhoukoudian Town

6.1.3 石灰岩
Limestone

周口店地区可供工业用的石灰岩分布较广,储量较丰富。周口店附近的龙骨山、太平山南坡、黄院等处的下奥陶统上部的马家沟组中有多层厚层状石灰岩,是开采水泥原料石灰岩的重要基地。

Abundant limestone for industrial use is widely distributed in Zhoukoudian area. In the areas near Zhoukoudian, Huangyuan, the southern slope of Taiping Mountain and Dragon Bone Hill, there is multistory thick-bedded limestone in the Majiagou Formation of the upper part of Lower Ordovician. These areas are important exploitation places for limestone which is a kind of cement raw materials.

周口店黄院水泥厂
Huangyuan Cemerit Plant in Zhoukoudian

周口店村东石灰岩采石场
Limestone Quarry in the east of Zhoukoudian Village

6.1.4 板岩
Slate

周口店地区可做建筑用石板的岩层主要是两个层位：一是三叠系双泉组中的凝灰质板岩；二是为晚元古景儿峪组上部的钙质板岩。北京奥运会鸟巢所用板岩就是景儿峪组板岩。

The Zhoukoudian area, the architectural slate is mainly distributed two Formations: Shuangquan Formation of Triassic and upper part of the Jingeryu Formation of late Proterozoic. The calcareous slates in the some of which are used in the "Nest" of Beijing Olympics.

拴马庄村石材加工厂
Stone Processing Plant in Shuanmazhuang Village

6.1.5 红柱石
Andalusite

红柱石为富铝硅酸盐矿物,具强耐火性,可作为高级耐火材料,为冶炼工业服务。

周口店地区红柱石的形成均与房山岩体有关,含红柱石的岩层较多。其中,下马岭组中的红柱石片岩,红柱石质纯,符合高级耐火材料要求。本溪组下部的红柱石角岩中不含炭者,亦是较好的耐火原料。

Andalusite is an aluminium silicate mineral. It has the fire-resistant speciality so that it can be used as refractory product in smelting industry.

There are many andalusite-bearing rocks in the Zhoukoudian area. The formation of andalusite here is associated with Fangshan pluton. The best quality andalusite is contained in the andalusite schist of the Xiamaling Formation, and conforms to national standards for superior fire-resistant material. The andalusite contained in the andalusite hornfels of the lower part of Benxi Formation is free of carbon, and is good refractory raw material too.

6.2 旅游资源
Tourism Resources

6.2.1 "北京猿人"遗址
"*Peking man*" site

　　"北京猿人"遗址——龙骨山是我国研究古人类学和古脊椎动物学的重要基地。在龙骨山的岩洞中保存有三"代"古人类化石及其生活遗迹：距今约60多万年的"北京猿人"、距今约10万年前的"新洞人"和距今约5万年的"山顶洞人"。

　　自我国著名的古人类学家裴文中在龙骨山猿人洞发现第一个完整的北京猿人头盖骨化石以来（1929），经过数十年的发掘工作，迄今已采集数百件古人类化石，据其可以复原出男、女、老、幼40多个猿人个体。与猿人化石同时出土的还有100多种脊椎动物化石、几万件古器以及猿人用火的遗迹。

　　龙骨山堪称古人类学的宝库，它对于研究人类的演化、人类社会的发展具有十分重要的科学意义。这些发现已陈列在"猿人博物馆"中，供游人参观。

The "*Peking man*" site, Dragon Bone Hill, is an important base in China study paleoanthropology and vertebratate paleontology. In the caves of Dragon Bone Hill, living heritage and fossils of three "generations" of ancient human have been discovered: "Peking Man" over 600 000 years ago, "New Cave Man" about 100 000 years ago and "Upper cave man" about 50 000 years ago.

Since the famous Chinese paleoanthropologist Pei Wenzhong discovered a complete skull of "*Peking man*" in Apeman Cave of Dragon Bone Hill (1929), archaeological excavation work has been going on for decades, hundreds of ancient human fossils have been unearthed. More than 40 individuals has been restored, including male and female, young and old. At the same time, more than 100 kinds of vertebrate fossils, tens of thousands of pieces of antiquities and the remains of ape-man using fire were discovered.

Dragon Bone Hill is a storehouse anciforent anthropology. It's of great significance for studying the human evolution and the development of human society. The findings are displayed in the "Ape Man Museum" which is open to the public.

6.2.2 十渡风景区
Shidu Scenic Spot

十渡位于北京房山区拒马河中上游。从千河口到十渡村,沿途在拒马河上要过桥渡水十次,"十渡"因此得名。

十渡是北京西山以岩溶峰林和深切河谷地貌为特色的自然风景区。这里河谷狭长蛇行,谷壁刀劈斧削,拒马河依山绕岭从西北至东南流贯全境,两侧山峰如塔似剑。东起千河口,西至大沙地,北到石人峰,南临笔架山,面积213平方千米,游览长度百余里。

十渡风景区内,无处不是奇峰秀水景象,因此,十渡被人们誉为"北方小桂林"、"百里画廊"。

Shidu is in Xishan area in the middle and upper reaches of the Juma River in Fangshan District of Beijing City. On the way from Qianhekou to Shidu Village, there were ten terries crossings along the Juma River such is the derivation of the Chinese name Shidu.

Shidu Natural Scenic Spot is featured by natural peak-forest karsts and deeply cutting river-valley which is long, narrow and like a fleeing serpent. The Juma River flows northwest to southeast through the area, with towering peaks and dangerous cliffs on either side. The total area of Shidu is 213 square kilometers, east to Qianhekou River, west to Dashadi, north to Shirenfeng and south to Bijiashan. The tour length is more than 100 li.

In Shidu Scenic Spot, grotesque peaks and beautiful waters are everywhere; hence Shidu is called "the Guilin in North China" and "Art Gallery".

6.2.3 石花洞景区
Shihua Cave

石花洞是中国北方四大名洞之一，是国家级风景名胜区及地学知识科普教育基地。2005年9月18日，获得"中国最佳溶洞奇观"称号。

石花洞发育在奥陶系马家沟组石灰岩中。洞体为多层多支的层楼式结构，洞内洞体分为上下七层，层次分布明显，洞穴沉积物分布密集、类型齐全、数量繁多。有滴水、流水、渗透水、停滞水和飞溅水五种沉积类型，四十多种沉积形态。洞中大量的月奶石为国内首次发现；石旗、石盾、石幔是中国洞穴沉积物的典型代表。石花洞中常年恒温13℃，四季如春，是一年四季旅游的好去处。

Shihua Cave is one of the four famous caves in north China. It is a national scenic spot and an educational base of geoscience. On September 18th, 2005, it was approved as "the most wonderful karst cave in China".

Shihua Cave developed in limestones of the Majiagou Formation of Ordovician. It has multi-layer and multi-branch, and can be divided into seven clearly demarcated layers. A great deal of sediments with different origins and scales are well developed in the cave, including more than 40 sedimentary states and five sedimentary types formed by drip water, artesian water, seepage water, stagnant water and splashed water. A large number of moonmilks were found here for the first time in China. There are typical cave sediments of China, such as stone shields, stone curtains and stone flags. With a constant temperature of 13℃, Shihua Cave is a very nice place for sightseeing at any time of year.

6.2.4 银狐洞景区
Silver Fox Cave

银狐石洞距北京 70 千米，是一个大型溶洞群，因发现罕见的形似狐狸的大型白色方解石晶体而得名。

银狐洞洞长超过 5 000 米，现已开发的近 3 000 米。其中主洞、支洞、水洞、旱洞纵横交错。银狐洞具有两大特点：一是洞里到处可见罕见的石菊花、晶花、石珍珠、石葡萄等景观，最令人叫绝的是长近两米、形似狐狸的大型方解石晶体，全身布满洁白剔透的毛绒针刺，为世界溶洞首次发现。二是在银狐洞最底层，深入地下 106 米的地下河。河水清澈见底，经国家级鉴定为优质天然矿泉水，因流经磁铁矿床成为天然磁化水，具有明显的消炎杀菌功能。银狐洞常年温度保持在 14～16℃左右。

Yinhu Cave (Silver Fox Cave) is 70 kilometers from Beijing city and is a scenic spot of water-eroded caves where researchers have found a fox-shaped snow-by yielding calcite crystal which is very rare and precious.

The cave is more than 5 000 meters long, of which nearly 3 000m has been explored. The cave is like a maze with the main cave, branch cave, water cave and dry cave intervened vertically and horizontally. It has two main features. First, a number of rare sights including stone chrysanthemum, crystal flower, stone pearl and stone grape can be seen everywhere. But the most astonishing one is a large calcite crystal which is nearly 2 meters long and shaped like a fox, covered by pure needle crystal clusters. Such a marvelous calcite crystal is the first that was ever found in the karst caves all over the world. Second, there are clear underground rivers at the bottom of the cave, in the depth of 160m. According to the state-level test, the water is high quality natural mineral water for it is naturally magnetized when flowing by magnetite deposit. The annual range of temperature in the cave is roughly from 14 to 16℃.

6.2.5 中国房山世界地质公园博物馆
Fangshan Global Geopark Museum of China

房山世界地质公园博物馆位于北京市房山区长沟镇，建设于 2009 年，是目前中国首座世界级的地质博物馆。博物馆占地面积 91.65 亩，主体建筑面积 10 000 平方米，馆内布展面积 5 800 平方米，分为标本陈列区、游客信息咨询服务中心、科普教育与科研交流中心、4D 动感电影厅、展品销售厅和办公接待中心六大区域，馆外设有 18 000 平方米的科普广场，馆内外展示区集中了岩石、动植物化石标本 100 余种类型，3 000 余例，模型装置 20 余处、多媒体演示 40 余处、模型场景 1 000 平方米。展陈内容以"地球演化、生物演化、人类进化"为主线，具有"科普教育、休闲旅游、地质研究、宣传展示、信息交流"五大功能，是中外游客认识房山的一张精美名片。

The Fangshan Global Geopark Museum is located in Changgou Town, Fangshan District, Beijing City. It was built in 2009 and it is the first world-class geological museum in China. The whole museum occupies 91.65 Chinese mu, while the main building occupies 10 000 square meters, a net exhibition area with 5 800 square meters. basically consists of six parts: Specimens Display Area, Visitors' Information and Service Center, Science Education and Exchange Center, 4D Film Room, Exhibits Salesroom, Office and Reception Center. The museum has a science plaza of 18 000 square meters. There are more than 3 000 specimens of fossils and rocks in the exhibition areas inside and outside the museum, including more than 100 different types. There are modeling devices in more than 20 different places, multimedia presentations in more than 40 different places, and scene models of 1 000 square meters. The exhibition is designed to hold. "Earth evolution, Bio-evolution and Human evolution" as the main line, with five main functions: "science education, leisure and tourism, geological research, exhibition and propaganda, information exchange". It is a fine visiting card which introduces Fangshan to Chinese and foreign tourists.

6.2.6 平西人民抗日斗争纪念馆
Memorial Museum of Pingxi People's war Against Janpanese Aggression

平西位于北京市房山区十渡村，总占地面积30余亩，建筑面积3 500平方米。平西抗日纪念馆以大量真实图片和史料反映抗日战争时期开创平西抗日根据地的肖克等一大批老将军的丰功伟绩，和牺牲在平西抗日战场上千千万万抗日先烈的英雄事迹，充分反映了当年30万平西人民的抗日斗争精神。是开展爱国主义教育的重要基地。

Memorial Museum of Pingxi People's War Against Janpanese Aggression is located in Shidu Village, Fangshan District, Beijing City, with a total area of more than 30 mu, and a floor area of 3 500 square meters. There are large amounts of pictures and historical documents which reflect the great contributions by many old generals, such as Xiao Ke, who established the Pingxi Anti-Japanese Base Area in the War of Resistance Against Japan, and heroic deeds by the revolutionary martyrs who sacrificed in Pingxi anti-Japanese battlefield. The exhibits fully reflect the struggle spirit of 300 000 Pingxi people united against the Japanese invasion. It is an important base for carrying out patriotism education.

6.3 水资源及环境保护
Water resources and environmental protection

6.3.1 周口河
Zhoukou River

图为经人工改造后的周口河河谷地貌景观（摄于周口店太平山西坡）

The figure shows the river-valley landform of the rebuilt Zhoukou River (The shot is taken at the west slope of Taiping Mountain in Zhoukoudian.)

6.3.2 牛口峪马刨泉
Mapao Spring in Niukouyu

马刨泉属碳酸岩岩溶裂隙水，赋存于本区奥陶系、寒武系灰岩含水层中，丰水期最大流量可达 1m³/s，枯水期不足 0.1m³/s，受季节影响泉水流量不稳定。

Mapao Spring belongs to fissure-karst water in carbonate rocks, hosts in the limestone aquifer of Ordovician and Cambrian. The spring flow is affected by the season with maximum flow in normal water season up to 1m³/s and less than 0.1m³/s in dry season.

6.3.3 房山区牛口峪污水处理厂
Niukouyu Sewage Treatment Plant of Fangshan District

位于房山区山顶庙。该厂 1995 年 1 月建立以来大大改善了周口店地区的用水、空气环境。现已将牛口峪水库改造成牛口峪生态中心，2007 年 7 月为环境科普示范基地，是大、中、小学生和游客学习环保知识、增强环保意识的好去处。

The plant is located in Shandingmiao of Fangshan District. Since its foundation in January 1995, the water environment and air quality in the Zhoukoudian area have been greatly improved. Niukouyu Reservoir has now been turned into the Niukouyu Ecology Center which became a demonstrative base of environmental science education in July 2007. It's a good place for students and visitors to raise consciousness and learn related knowledge about environmental protection.

6.3.4 牛口峪水库
Niukouyu Reservoir

经综合治理后的牛口峪水库,现已成为当地居民水上娱乐休闲场所。

After comprehensive treatment, the Niukouyu reservoir has become a place of entertainment and leisure for local residents.